In Search

Birds in

Wales

Brian O'Shea

Front Cover-Red Kite from an original water colour by Owain Williams
Rear Cover-Peregrine Falcon from an original water colour by Owain Williams

Published by - 'Skylark Books', 'Ty-Nant', Dyffryn, Llanilar, Aberystwyth SY23 4PF

To order – e-mail: enquiries@skylarkbooks.ndo.co.uk

ISBN No. 0-9538115-0-6

Printed by – Cambrian Printers, Llanbadarn Road, Aberystwyth SY23 3TN

CONTENTS

ACKNOWLEDGEMENTS

This book owes much to the support of relatives and friends; to my son-in-laws Paul and Cherif for launching me into the computer age, to my daughter Wendy who has taken time out from her own studies to offer invaluable guidance in setting out the text, and to my wife Valerie who typed the original script and who has driven the length and breadth of Wales with me re-examining bird sites. I am particularly indebted to Kate and Robert Cooper, Kate for the many hours spent over a period of several months teaching me the rudiments of using a computer, thus enabling me to get the book under way, and to Rob for his meticulous work in proof reading the text. Once more, I have called upon the skills of my long time birdwatching companion John Green to contribute to the literary style, and John has also provided most of the black and white drawings. I am especially grateful to Peter Davis, former ornithologist to the Nature Conservancy Council and County Recorder for Ceredigion, for checking the content of the script and making many useful suggestions and corrections. Although now retired, Peter is still very active in the ongoing work of protecting Kites in Wales.

A book like *In Search of Birds in Wales* cannot be produced entirely on the experiences of one person alone. A significant proportion of the information offered in this book is due to the many researchers and ordinary fieldworkers and members of organisations like the RSPB, the British Trust for Ornithology, and the Welsh Ornithological Society. Although most of them are not known to me personally, they have, in a sense, all contributed to this work. In particular, I should like to acknowledge the W. O. S. and its members who submitted their records, the county recorders who collated them, and the unpaid staff who have turned out reports annually since the inaugural meeting of the society at Aberystwyth in 1988. To all of them I owe my thanks.

IN SEARCH OF BIRDS IN WALES

INTRODUCTION

In Search of Birds takes you on journeys into the Welsh countryside, from lanes and woods to the highest moors and mountains. The many varied habitats are explored in turn for the birds which can be discovered within them. Its aim is to be descriptive and readable whilst at the same time providing comprehensive details on bird distribution and populations, behaviour, habitat choice, recent species history and conservation issues.

The pleasures of birdwatching in Wales are enhanced by the superb natural scenery and by the many opportunities available to wander its remote uplands, rugged coasts and verdant valleys. In such landscapes you can seek out those 'special' birds for which Wales is renowned, notably its birds of prey, seabirds, and the summer migrants which inhabit its upland woods. Foremost among these birds of course, is the Red Kite, a star bird not only for its size and grace but also for its historic struggle for survival during the twentieth century. Then there is the majestic Buzzard soaring over the hills and valleys, as familiar a part of the rural scene as the sheep and the oak woods over which it flies. The Buzzard and the Raven prevailed in the remoteness of the Welsh hills during times of persecution and today both species are still found in higher concentrations here than anywhere else in Britain. The cliffs of Wales sustain the only populations of the Chough on the British mainland, while the country can also claim a good proportion of those bold and breathtaking avian predators, the Peregrine Falcon and the Goshawk. The rugged western cliffs, and especially the 'bird city' islands off the Pembrokeshire coast support enormous numbers of seabirds such as Gannets, auks, Kittiwakes and Manx Shearwaters.

The deciduous woodlands together with the bracken-covered hills or boulder strewn slopes are the classic habitats in Wales. Between them they claim high densities of attractive summer visitors: Pied Flycatchers, Redstarts and Wood Warblers frequent the woods, while Whinchats, Tree Pipits and Wheatears inhabit the open hillsides. There are fascinating though often elusive and declining birds to search for on the hills including the Black and Red Grouse, the Dunlin, Merlin and Ring Ouzel. The coniferous forests, which have proliferated over the uplands in the past 50 years, have brought with them a crop of new exciting species. The Goshawk, Crossbill and Siskin all owe their current status in Wales to the vast conifer plantations. During the same period, Goosanders and Red-breasted Mergansers arrived from the north to join the traditional Dippers and Grey Wagtails on our fast flowing rivers.

Many of the characteristic Welsh breeding birds mentioned above are scarce or absent in lowland English counties to the east but although these species claim a special link with Wales, they represent only a small proportion of the total. There is much more to see in the varied and dramatic landscapes of this country. More than 150 species of birds breed annually and another 20 have done so at various times. All of these together with a further 90 or 100 winter visitors, passage migrants, and occasional visitors, are discussed in the pages of this book. Common species are described along with those rarities whose anticipation and discovery infuse a sense of adventure and excitement into any birdwatching outing.

A 'birds by habitat' method of organising the book has been adopted for two main reasons: firstly to describe birds in the context of the Welsh countryside, and secondly to focus on the species that are most likely to be found in particular habitats. To the less experienced birdwatcher this should be an invaluable guide to identification. Of course, several habitats are usually encountered during even a short walk and then there will be a corresponding increase in the number of species which can be expected. It should also be remembered that many common farmland species described in Chapter One, may occur in any habitat where there are trees or bushes. The bird lists in small print and notes at the end of each chapter are also designed to help the reader to target the species which are most likely to be encountered in each habitat.

Whilst the first six chapters link birds closely with their breeding habitats, Chapter Seven 'Wales in Winter' is treated rather differently because at this time of the year the relationship between birds and their habitats changes. Most upland birds, for instance, desert the high ground and seek shelter in low-lying places. Finches, sparrows, tits and larks leave their home-based farms and obtain food, warmth and protection from predators in the security of flocks. The majority of our winter visitors come from northerly latitudes and wildfowl and waders figure highly among them, so at this time of the year coastal locations, lakes and marshes are especially important. Wintering birds are dealt with therefore under one chapter heading, starting with those which frequent watery places and continuing later in the chapter with species which are more likely to be met with in dryer habitats. It should be noted that winter is defined very loosely to include the autumn months from late September.

Passage migrants which travel through Wales on their way to winter in warmer climates, or return in spring to their northern breeding grounds are dealt with in several chapters. A significant number of these species are waders and seabirds. I have chosen to include a list of migrants at the end of Chapter Seven, noting in which months each species is most likely to be observed. Waders such as the Greenshank, Little Stint, and Whimbrel are typical of the birds eagerly awaited in waterside habitats, but much rarer species are possible. Timing is crucial. Generally speaking August, September, October, April and May are the most important months. Locations almost devoid of bird life at other times of the year are now keenly watched for the arrival of scarce and interesting species.

The last chapter is dedicated solely to birds of prey, since this fascinating group is so well represented in terms of both numbers and variety of species in Wales. They are also difficult to 'pin down' to specific haunts. All feature in the earlier chapters dealing with their preferred nesting habitats, but for many raptors the sky is their domain when hunting, as they scour the land widely in search of prey. The final section of the text provides notes on the range and status of each species in Wales. All species are included except rare vagrants usually caught in mist nets, or which turn up once or twice on an offshore island during the course of a decade.

Although many characteristic species may be encountered widely, there are numerous locations which are particularly fruitful for bird watching. I have listed nearly 60 of these in the opening section of the book but this is by no means an exhaustive list. Estuaries, headlands, and scarce specialised habitats such as fresh water marshes and lakes figure highly among these favoured locations. A high proportion of these sites are owned or managed by conservation organisations such as the Wildlife Trusts, The Countryside Council for Wales, The National Trust and the RSPB. The Wildlife Trust West Wales alone, for example, manages more than 60 sites in the counties of Carmarthen, Ceredigion and Pembrokeshire. Plants, insects and other wildlife benefit from habitat protection and management as well as birds. Birdwatching at some of these prime sites is likely to lead you into contact with other birdwatchers, especially where the provision of hides brings people close together. This is the opportunity to learn or share skills in identifying birds, exchange information or maybe join in the work of the organisation.

We can all be thankful for the activities of conservation groups because not all is well with wildlife in the Welsh countryside. As elsewhere, bird populations generally seem to be on the decline. Well loved species like the Skylark and Lapwing, the Song Thrush and Yellowhammer, are all disappearing rapidly from farmland fields. On the upland moors our precious populations of Red and Black Grouse and breeding waders such as Dunlin, Golden Plover and even the Curlew, are suffering from the effects of overgrazing by high densities of sheep, drainage and the degradation and depletion of heather habitat.

The counties of Wales and their borders were reorganised in 1974 and again in 1996 and each time the names and county boundaries were substantially changed, so that people could be forgiven for wondering which county they actually lived in! The town of Conway for instance, started in Caenarfonshire, became part of the larger county of Gwynedd in 1974, and from 1996 has been situated in the new county of Conwy. Similarly, the residents of Pontypool have, during the same period, experienced life in Monmouthshire, Gwent, and now Torfaen. By and large, and at risk of seeming outdated, we stick to the old but easily recognised counties and their boundaries! For the purposes of the book, we frequently consider birds on a regional rather than a county basis, partly because these are readily understood despite a lack of definitive boundaries, and partly because these divisions seem to be less taxing on the readers' knowledge of geography and local government. Our regional maps devised purely for ornithological purposes contain numbered sites which correspond to those listed in the accompanying text.

In Search of Birds is designed to inform the reader which birds occur regularly in Wales and where they may be seen. We consider the status of the various species not only in their Welsh context but sometimes in a wider British or European one, since none are totally confined to the principality. Tips based on nearly 50 years experience of bird watching are given on the matter of identifying birds that we hope will prove beneficial when combined with the use of a standard illustrated guide.

The overall aims of the book are to describe birds in the setting of the Welsh countryside, offer a range of useful and interesting information, and provide assistance in the business of finding and identifying birds. We hope this book will help you not only to locate your birds more effectively but identify them more easily when you find them!

THE REGIONS OF WALES

This section of the book is designed to focus on the geographical regions of Wales; the key mountain and upland areas, major rivers and lakes and prime bird watching locations.

Many of Wales' most notable birds are to be found over wide areas of the countryside but some species require highly specialised habitat on islands, steep cliff faces, freshwater lakes or marshes. Some locations are especially good for birds on migration or wintering wildfowl and waders. Others offer refuge for rare or local breeding birds such as Reed Warbler, Little Tern and Puffin.

Bird locations of special interest, many of them reserves under the management or ownership of conservation organisations, are marked on the maps and listed below under each region. This is not a book in the 'Where to Watch Birds' series and the annotated remarks do not give detailed directions of how to get to the 60 or so places described. A decent modern map, however, and if necessary a word with a local resident (or a phone call to the agency which manages the site), should enable almost all of them to be found with little trouble. Only species of particular interest to be expected at each site or district are mentioned, and in many cases the principal locations and their birds are discussed further in the following chapters.

At some reserves there may be restrictions on the days of the week you may visit and the parts of the site where entry is allowed. A permit may be required

A rural scene in Wales

or an admission charge may be made, so it is better to make enquiries first before embarking upon a long and fruitless journey.

Although season and location are obviously very important, time of day and weather conditions will also drastically influence the number and variety of birds which can be expected. Passerines such as tits, warblers and finches are most active until about mid morning, and to an extent, again during late afternoon or early evening. The hours between about 11.00a.m. and 3.00p.m. are usually unrewarding for observing these kinds of birds. Birds of prey on the other hand, like to soar on the rising thermals during the warmer parts of the day. Harriers and owls are best looked for some time before dusk while sea-watching is most productive just after dawn. A study of the tides will also pay dividends: shoreline waders for example are most easily observed when the area of exposed mud or sand is small, preferably as the tide is just receding. Understandably, birds are no keener on wet and windy conditions than humans are, but for them it may be a matter of survival. Energy sapping cold winds deplete the strength of small birds in particular, causing them to keep in close cover. Onshore gales however, may be good news for birders since they may bring with them a crop of oceanic seabirds or a landfall of passing migrants.

Region One
NORTH WALES

There are no clear cut boundary lines between North and South Wales and precise distinctions are even harder to make if Mid Wales is classed as a separate region, as it is in the pages of this book. For many South Walians travelling north along the A487, North Wales begins just across the bridge at Machynlleth, but for our purposes the maps below will be used to describe the three regions ornithologically!

Think of North Wales and you think of mountains and rugged peaks. The mountains of this region, the grassy upland Carneddau, the craggy Glyder, and the towering Snowdon can claim between them all of the summits over 3000 feet in the whole of Wales. From this lofty terrain the rivers draining the mountains descend through rocky gorges and steep sided valleys strewn with boulders or hung with oak and pine. Here there are still red squirrels to be seen and this is one of the best parts of Wales to encounter Peregrine, Choughs and Ring Ouzels. East of the Conway the hills are gentler and the rainfall lower and rolling farmland replaces rugged mountains. In parts of Denbighshire these hills are bedecked with rich heather moorland where the Merlin is to be found nesting and Grouse can still be encountered, though in declining numbers, on the Berwyn Mountains. Keepered moorlands such as those by Ruabon near Llangollen are fewer than in former times, when large bags of Grouse were taken each season. The north-east of Wales near the River Dee has become heavily industrialised but this low lying flood plain in Flintshire holds our remnant population of Corn Buntings.

Travelling west from Rhyl, the route becomes increasingly picturesque as you approach Llandudno and Conway. The coastal plain of North Wales reaches its narrowest point near Penmaenmawr, where the mountains reach almost to the shore. Here the road is tunnelled through the hills in places and the widened trunk road is forced for lack of space almost onto the beach. There are numerous coastal sites for bird watching in North Wales such as Traeth Lafan (Lavan Sands), Conway and of course the Dee on the eastern border of Wales. Great Orme's Head shelters colonies of seabirds and regularly turns up interesting passage migrants. Anglesey has tern colonies and Fulmars, auks and gulls on its higher cliffs and migrating and wintering birds in Red Wharf Bay and off Point Lynas. Bardsey Island in the west has recorded more species of rare visitors than anywhere else in the whole of Wales.

From a position near the Menai Straits, the low lying Anglesey presents a sharp contrast to the line of mountains just across the water on the Welsh mainland. Yet with its beautiful wild coastline, dunes, lakes and marshes, Anglesey has a wealth of wildlife. In the north-west of Caernarfonshire, the Lleyn Peninsula is characterised by a long unspoiled coastline reminiscent of Pembrokeshire. As many holiday-makers have discovered, some of the beaches and coastal scenery in this area rank among the best in Britain. Here and on neighbouring Bardsey, the Chough is found in good numbers and Peregrine eyries and seabird colonies are scattered all along the coast.

Meirionnydd vies with Breconshire as the district which has the highest proportion of land in Wales over 500 metres and it is one of the best parts of the country for several species including Black Grouse, Merlin, Nightjar and Peregrine. The Snowdonia National Park extends to the Aran mountains and Cader Idris which forms a backdrop to the picturesque Mawddach estuary, considered by many people to be one of the most beautiful places in Wales.

Locations of Special Interest

1. SHOTTON POOLS
Habitat – Pools, reedbeds, rough ground.
Birds – Waterbirds especially duck. Sometimes Jack Snipe in winter. Waders and other scarce birds on passage. Common Terns, Yellow Wagtail, Reed Warblers, Whinchat, Redshank and possibly Corn Bunting breed.
Access – Turn right at end of A548 onto A550, then left into steel works. Permit required from British Steel (write for details).

2. FLINT/DEESIDE SHORE
Habitat – Estuarine marsh and river, mudflats.
Bird – Waders and duck, including huge numbers of Shelduck, Knot, Dunlin and Oystercatchers in winter.
Access – Coastal footpaths close to the castle in Flint. Free parking, turn left after the lights heading south on A548. Reserve at Connah's Quay by Deeside power station (Powergen). Access by permit only, turn left on A548 at roundabout 1mile SE of Flint.

3. POINT 0F AYR/GRONANT
Habitat – Shallow Bay, saltmarsh, mudflats.
Birds – Wildfowl and waders. Wintering predators, sometimes rare

passerines e.g. Shorelark, Snow bunting, Twite. Good sea-watching. Little terns breed at Gronant (no public access to site available).

Access – Turn off A548 two miles east of Prestatyn (after Gronant) signposted Talacre, to RSPB hide overlooking estuary.

4. CONWAY ESTUARY

Habitat – Estuary, freshwater lakes.

Birds – Waders, wildfowl, Ospreys on migration. Little Ringed Plovers breed.

Access – Access most days (not Tuesdays) to this RSPB reserve. Turn left off A50 just before Conway bridge travelling west.

5. GREAT ORME

Habitat – Limestone headlands, grass slopes and cliffs.

Birds – Seabird colonies, scarce migrants.

Access – Via the Marine Drive from Llandudno Pier at the west end of the promenade.

6. TRAETH LAFAN

Habitat – Extensive exposed sand and mudflats.

Birds – Wintering and passage waders, sea duck and Great Crested Grebes.

Access – Shoreline access between Bangor and Llanfairfechan. Coast path for several miles, best access Port Penrhyn, Llanfairfechan and Morfa Aber - turn off A55 eastbound signposted Abergwyngregin.

7. FEDR LAWR

Habitat – Rocky shore, cliffs and scrubland.

Birds – Only site of breeding Black Guillemots, also Fulmars breed. Scrub birds may include Grasshopper Warblers, Linnets and Yellowhammers.

Access – Minor access lanes near village of Llangoed to this National Trust site.

8. POINT LYNAS

Habitat – Rocky headland.

Birds – Migrating and passing Manx Shearwaters, Gannets and terns, also divers and sea duck. Rarer species regular such as Leach's Petrel and Sooty Shearwater.

Access – Take unclassified road off the A5025 to the east of Amlwch signposted Llanelian. Drive 1 mile beyond village, above harbour to National Trust headland.

9. CEMLYN BAY

Habitat – Lagoons and shingle and rocky shore.

Birds – Tern colonies, visiting duck and waders. Seabirds offshore.

Access – Turn off A5025 at Tregele on unclassified roads and use car park either Traeth Cemlyn or Bryn Aber. National Trust site managed by North Wales Wildlife Trust.

10. SOUTH STACK
Habitat – Steep cliffs and heather covered heaths.
Birds – Auks including Puffin. Chough and Peregrine. Scarce seabirds, migrants and accidentals. Hen Harrier and Merlin in winter.
Access – Turn off B4545 or minor roads out of Holyhead (RSPB).

11. MALLTRAETH
Habitat – Estuary, pine woodlands, pools and water meadow.
Birds – Wildfowl, waders, woodland birds.
Access – A4080 between Newborough and Rhosneigr. Mostly Forest Enterprise, also RSPB reserve at coast.

12. NEWBOROUGH WARREN
Habitat – Extensive dunes, dune grassland and pine forest and pools. Rocky promontory and sandy beach.
Birds – Wildfowl, waders, Short-eared Owl and Hen Harrier.
Access – From A4080 near Newborough. Footpaths to shore. Forest Enterprise road west from village centre.

13. VALLEY LAKES
Habitat – Pools, rough pasture, gorse, reedbeds.
Birds – Pochard, Shoveler, Gadwall, Tufted Duck and Ruddy Duck breed.
Access – Turn off A5 after Bryngwran, signposted R.A.F.Valley. There is RSPB pathway on right after first lake.

14. LLYN ALAW
Habitat – Large freshwater lake.
Birds – Highest population of wildfowl inland in Wales.
Access – Turn right, lake signposted off A5. Small car parks at termination of unclassified roads. Car park, two hides, and visitor centre at east end of lake (Welsh Water reserve).

15. BARDSEY ISLAND (YNYS ENLLI)
Habitat – Island two miles offshore.
Birds – Resident seabirds, migrants including rarities, Choughs.
Access – Boat from Port Meudwy, near Aberdaron, March-October. (accommodation for birdwatchers - prior booking required).

16. GLAS LLYN
Habitat – (a) Flood meadows, river.
(b) Estuarine mudflats.
Birds – Wildfowl including Whooper Swans, waders on the seaward side of the the road and railway.
Access – Footpath along the causeway embankment starting near miniature railway station at Porth Madog permits views of both habitats.

17. TRAWSFYNYDD
Habitat – Lake and marshy fields.
Birds – Waders, gulls, breeding waders and grebes.
Access – Non classified road south of the lake.

12

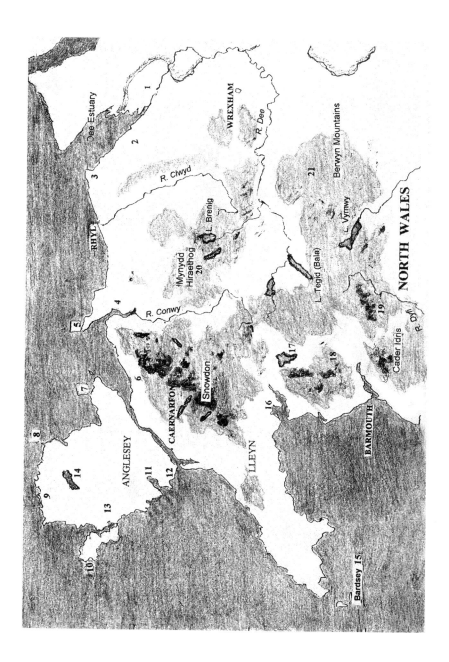

NORTH WALES

Dee Estuary
WREXHAM
R. Dee
Berwyn Mountains
R. Clwyd
21
L. Vyrnwy
RHYL
L. Brenig
Mynydd
Hiraethog
20
L. Tegid (Bala)
R. Conwy
R. Dyfi
Cader Idris
19
Snowdon
CAERNARFON
17
18
ANGLESEY
16
LLEYN
14
BARMOUTH
11
12
13
10
Bardsey 15

18. COED Y BRENIN
Habitat – Extensive coniferous forest and riverside habitat.
Birds – Siskin, Crossbill, possibly still Black Grouse.
Access – Several walks from visitor centre approached from A470 north of the village of Ganllwyd.

19. DYFI FOREST
Habitat – Extensive coniferous forest, open ground and deciduous woodland.
Birds – Species include Nightjar, Siskin, Crossbill and Peregrine.
Access – Unclassified road between Aberangell and Corris.

20. MYNYDD HIRAETHOG
Habitat – Lakes, heather moors, bog and coniferous forest.
Birds – Great Crested Grebes on Llyn Brenig, Red Grouse and moorland waders including Golden Plover and Dunlin breed. Wildfowl on lakes in winter.
Access – Via the A543 and unclassified roads.

21. BERWYN MOUNTAINS
Habitat – Heather moorland.
Birds – Red Grouse, Black Grouse, Merlin, Hen Harrier.
Access – Many access points to perimeter of moors.

Region Two
MID WALES

The boundaries of Mid Wales are even harder to define than those of North Wales. For the purposes of this book we have taken them to extend from the Cader Idris and Lake Vyrnwy in the north, to Cardigan and Brecon in the south. Much of this region is virtually unpopulated save for small market towns, interspersed with villages and hill farms. Powys sometimes known as the green desert has less than 100,000 inhabitants, while Ceredigion has about 70,000 people. Newtown, Aberystwyth, Welshpool, Llandrindod Wells, Builth and Brecon are the largest towns and none has a population exceeding 12,000. Towns like Machynlleth and Rhayader are set amidst outstanding countryside typified by bracken-covered hillsides, heather slopes topped by rocky crags, deep green valleys and deciduous woodlands. The crags above the Elan Valley lakes still provide haunts for breeding Peregrine Falcons and Ring Ouzels despite the popularity of the area with visitors.

With the exception of the Cader Idris and Arans in the north, the highest hills lie along the central spine of the Cambrians where they reach a maximum of about 2500ft. Mid Wales is a land of tall hills and wooded valleys rather than mountains, and these provide ideal homes for the Buzzard, Redstart, Pied Flycatcher and Red Kite. In parts, the highest Cambrian uplands cradle a soggy wilderness for some of our rare waders though the moors are threatened by the encroachment of ever increasing numbers of sheep. Many of Wales' new plantations sprawl on these uplands where they give shelter to Crossbill, Siskin and Goshawk. There are some splendid coastal habitats for birds, most notably the Dyfi estuary which is now the Welsh stronghold for the Redshank and Lapwing. The Dyfi provides one of the best prospects of coming across a rarity in Mid Wales, and the coast off Borth and the estuary support Red-throated Divers, duck and numerous species of wader at different seasons. Aberdysynni to the north has Eider and Scoter throughout the year and large numbers of roosting Mergansers, Shelduck and other wildfowl.

To the far south of the region, the Teifi estuary is a stopping place for wintering birds and waders on migration. The hinterland just beyond the town of Cardigan offers the Teifi marshes with their deep reed beds which attract many species of warbler including, until recently, Wales' main concentration of the rare Cetti's Warbler. Tregaron Bog or Cors Caron, is a wetland of a different kind. This is the largest raised peat bog in southern Britain and extends for some four miles north of Tregaron. Most of it is managed by The Countryside Council for Wales and is covered by a combination of heather moor, rough grasses, willow carr and birch. The bog is mainly of botanical interest but Kites are almost guaranteed and Hen Harriers are a major attraction for birders visiting the reserve in winter. Cors Fochno at Borth is also an extensive area of bog and is good for a variety of birds of prey, especially in winter.

Mid Wales contains several of the largest reservoirs in Wales including Llyn Brianne, Nant y Moch, Llyn Clywedog and the Elan Valley lakes. In the main, it must be said, bird watching prospects are limited on these waters which are deep, acidic and as a rule have little waterside vegetation. The region also boasts some of Wales' longest rivers, although the Severn and Wye, which both have their sources in Plynlimon, flow through the border counties into England before returning to finally empty into the Bristol Channel. Near to the border these two rivers course through fertile landscapes which attract species like the Yellow Wagtail and Tree Sparrow which are scarce elsewhere in Wales. The Ceredigion coastline to the north of Cardigan provides wonderful seascapes and is home to several mixed colonies of seabirds, especially at New Quay Head, Llangranog and The Wildlife Trust West Wales reserve at Penderi. Peregrine and Chough can be seen during a good day's walk almost anywhere along this stretch of coast.

Locations of special interest

1. ABERDYSYNNI
Habitat – River, lagoon, damp meadow, shingle/dunes and shore.
Birds – Sea duck, other wildfowl, waders, visiting migrants and occasional rarities.
Access – Via unclassified beach road north of Tywyn.

2. YNYS HIR
Habitat – Extensive RSPB reserve; mixed habitat includes saltings, pools, marsh, woodland and farmland.
Birds – Large range of species possible. Notable for wintering Greenland White Fronted and a small number of Barnacle Geese.
Access – Leave A487 at Furnace (Reserve sign-posted on left).

3. BORTH/YNYSLAS

Habitat – Dunes, beach, shallow sea, estuary.

Birds – Wintering divers, scoters and Great Crested Grebes, passage waders and terns, occasional rarities. A good variety of birds at all seasons.

Access – Access from coast road at Borth, car park at Ynyslas.

4. CORS FOCHNO (BORTH BOG)

Habitat – A huge area of bog, partly heather covered, with scattered reed-beds and intersected by tidal River Leri.

Birds – Reed Warblers, Water Rail and Snipe breed. Wintering birds of prey. Migrants and visiting rarities.

Access – Through golf course direct from beach road or turning in Borth towards animalarium. A National Nature Reserve run by the Countryside Council for Wales.

5. ABERYSTWYTH

Habitat – Rocky shore, harbour, hillside and cliff (Constitution Hill).

Birds – Purple Sandpipers in winter, sometimes scarce gulls. Peregrine and Chough on cliffs to the north, also Stonechats.

Access – Footpaths and promenade provide ready access.

6. RHEIDOL AND YSTWYTH VALLEYS

Habitat – Typical fast flowing rivers with farmland, deciduous woodlands, (especially sessile oak and beech) and coniferous forest. Moorland at higher reaches of these rivers.

Birds – Good variety, especially birds of prey, including Kite, Buzzard Goshawk and Peregrine.

7. PENDERI (MONKS CAVE)

Habitat – Cliffs with gorse and bracken covered hills, coastal sessile oak wood.

Birds – Chough, Peregrine, Stonechat. Breeding seabirds include Fulmar, Cormorant, Shag and Herring Gull (no auks).

Access – Pull in on A487 about one mile south of Blaenplwyf. Take track to coast path and turn south. The site is owned by the Wildlife Trust West Wales.

8. NEW QUAY HEAD

Habitat – Headland and steep cliffs.

Birds – Best place for breeding seabirds in Ceredigion, especially Kittiwakes, Guillemots and Razorbills (no Puffins). Also Choughs and Peregrines.

Access – Footpath from New Quay towards Bird Rock (head).

9. LLANGRANOG / YNYS LOCHTYN

Habitat – Cliffs, bracken and gorse hill slopes.

Birds – Nesting sea birds include Razorbills, Fulmars and Kittiwakes. Also Choughs and Stonechat.

Access – Coast paths both sides of the village sea front.

10. TEIFI MARSHES

Habitat – Varied - woodland, gorse, scrub, tidal river and saltmarsh.
Freshwater pools and extensive reed-beds are most important.

Birds – Good variety, especially noted for breeding Water Rail, Reed, Sedge
and Cetti's Warblers.

Access – Turn left 3 miles from Cardigan on the A478, signposted Cilgerran.
Turn left again after 1/2 mile and take track to car park Visitor
centre. Charge made. Owned and managed by the Wildlife Trust
West Wales.

11. PLYNLIMON (PUMLUMON)

Habitat – High moorland vegetation, scree.

Birds – Dotterel in May or Aug-Sept. A few Golden Plover, Red Grouse
and Ring Ouzel breed.

Access – Easiest from the private car park at Eisteddfa Gurig on the A44,
or from Glaslyn Nature Reserve near Dylife (off B4518).

12. ELAN VALLEY

Habitat – Extensive area with five reservoirs, woodlands, crags and heather
covered or grassy moors with bogland.

Birds – Peregrine Falcon and Ring Ouzel on the crags; Redstart,
Pied Flycatcher and all three woodpeckers in woodland. Golden
Plover, Snipe and Dunlin on moors. Goosander on lakes.

Access – The reservoirs are run by Welsh Water. The RSPB manage an area of
woodland and cliff slopes to the south of Elan Village.
Free access along many tracks and paths.

13. CORS CARON (TREGARON BOG)

Habitat – Huge area of raised bog covered with heather, grass and willow
carr. A few pools, area is traversed by Teifi River.

Birds – Birds of prey, especially Kites and Hen Harriers (in winter).
Occasional rarities. Grasshopper Warblers, Snipe, Redshank and
Water Rail all breed.

Access – Pull in by the B4343. Access along former railway track. Permit
required for tracks within the reserve. A large part is a National
Nature Reserve managed by the Countryside Council for Wales.
Kites are fed at Pont Einon - afternoons, Oct-March.

14. MOUNTAIN ROAD (Tregaron to Abergwesyn and Llanwrtyd Wells via Irfon Forest or Tregaron to Rhandirmwyn via Llyn Brianne).

Habitat – Wild hills, cliffs, moors, rivers - upper Tywi and Irfon. Deciduous
and conifer forest.

Birds – Kite, Peregrine, Merlin, Goshawk, Crossbill, Siskin, Common
Sandpiper, Goosander.

15. LAKE VYRNWY (and surrounding moors)

Habitat – Lake, forest and heather moors.

Birds – Great Crested Grebe, Tufted Duck, Goosander, Heron, Kingfisher

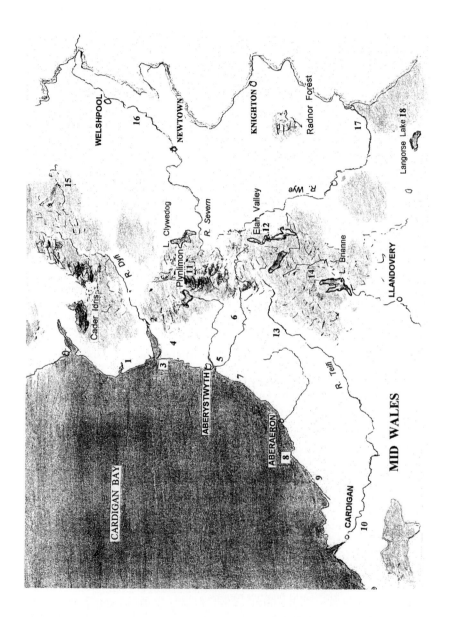

MID WALES

CARDIGAN BAY

WELSHPOOL
16

NEWTOWN

KNIGHTON
Radnor Forest
17

Langorse Lake 18

R. Wye

15

L. Clywedog
R. Severn

Elan Valley
12

R. Dyfi

Plynlimon
L. Vyrnwy
11

Cader Idris

L. Brianne
14

LLANDOVERY

1
2
4
5
6
3
13
7

ABERYSTWYTH

ABERAERON
8

R. Teifi

9

CARDIGAN
10

and Dipper breed on or near the lake and river. Moors notable for Hen Harrier, Merlin and Red and Black Grouse.

Access – RSPB reserve. Follow local instructions regarding public pathways. Road access through moors and around lake.

16. SEVERN VALLEY (Llanidloes to Welshpool)

Habitat – Fertile meadows and riverside.

Birds – Most species associated with Welsh rivers breed, including Dipper, Kingfisher, Common Sandpiper, Little Ringed Plover and Yellow Wagtails.

Access – Wildfowl etc. at Montgomery Wildfowl Trust reserve (with hide) by A483 - turn towards Powys Castle and use trust car park on right. For river birds, visit the trust's reserve at Dolydd Hafren, about two miles from Fordham.

17. GLASBURY ON WYE (and Llyswen / Boughrood)

Habitat – Fertile river valley, mixed crops and river shingle banks.

Birds – Noted for breeding Yellow Wagtail, Tree Sparrow and Little Ringed Plover. Wild swans in winter.

18. LLANGORSE LAKE

Habitat – Shallow, reed lined freshwater lake.

Birds – Reed Warbler, Yellow Wagtail, Water Rail and Great Crested Grebes all breed. Various wildfowl especially in winter. Passage migrants, sometimes rarities e.g. Bittern.

Access – Via access road to caravan park near Llangorse village or minor road to Llangasty -Talyllyn church on the south shore. (Footpath goes round part of the lake).

Region Three
SOUTH WALES

South Wales has hilly landscapes and deciduous woodlands which are typical of so many parts of Wales but generally there is more fertile lower ground where farms are more productive and fields under the plough are not unusual. This part of Wales is the most prosperous and contains a good proportion of the country's three million population. Cardiff the capital, Swansea and Newport are the biggest urban centres while the former coal mining valleys hold populous towns such as Merthyr Tydfil, Pontypridd and Neath. The outward scars of industrialism in the valleys have partly been covered over with landscaping but the social scars of unemployment remain. The bird life in the hills includes some of the typical upland species such as Buzzard, Wheatear, Raven, Peregrine, Whinchat and the declining Ring Ouzel. Hillsides clothed in bracken or mixed woodlands and capped with heather reach almost to the suburbs of Cardiff and harbour a rich variety of birdlife. Buzzards may be seen soaring above the Asda car park while Nightjars churr on some of the Glamorgan hills in summer. To the north of the 'Valleys' is the Brecon Beacons National Park, which rises to nearly 3000ft. The sandstone mountains here have smooth sweeping contours so different from the rugged hills of Snowdonia. On both the Beacons and The Black Mountain of South Carmarthenshire there are still one or two pairs of Red Grouse, Merlin and Golden Plover, but these species are becoming increasingly hard to find. The similarly named Black Mountains, to the north of Abergavenny, hold more promise and are worth exploring for typical upland species.

To the east of the region, Gwent (now mostly Monmouthshire) is a warmer and drier county attracting such species as Hawfinch and Turtle Dove which are rarely found elsewhere in Wales. This county was the last refuge of the Nightingale in Wales, but unfortunately the species' range has contracted eastwards and its musical notes are no longer to be heard regularly in its woodland thickets. The Hobby though, has been expanding westwards and in the past decade or two Gwent has become its first Welsh stronghold. It is to be hoped that this exciting species will soon become established in other parts of Wales. The urban areas of south-east Wales are not without their birdlife. Coots, Great Crested Grebes and even Moorhens, a bird surprisingly scarce in some areas, are to be seen on lakes and reservoirs around the capital. Interesting migrant waders or Black Terns may also pass through, while in winter, rarities like Smew or Bittern may turn up. The parks attract many species: Herons, Green Woodpeckers, Mistle Thrushes and Sparrowhawks to name just a few, while the Gwent coastline to the east can be searched for waders. The arable land of the Vale of Glamorgan is good for larks and Stock Doves, while Kestrels are regularly seen hunting over the motorway.

South Wales is an excellent area for winter bird-watching. The Glamorgans are probably the best counties in Wales to find Smew or rare gulls, while at critical months Pomarine Skuas are regularly recorded off Laverock Point. The prize-winning spot for rare gulls in Wales must be Black Pill at Swansea, where Little Gull, Mediterannean Gull and Glaucous Gull are not unusual. The star attraction though, is the very rare Ring-billed Gull which has been frequently observed here at all times of the year. The Burry Inlet and Cleddau estuary, a unique and picturesque network of wooded bays, coves and muddy creeks, attract substantial numbers of wintering wildfowl and waders such as Knots, Dunlin, Grey Plovers and Spotted Redshank.

Offshore, Carmarthen Bay holds huge numbers of Scoters, Red-throated Divers and smaller numbers of scarcer species in winter. This area of course, was badly affected by the huge oil spill from the stricken tanker 'The Sea Empress' in February 1997 but is slowly recovering. The true cost to wildlife is yet to be discovered but we can be thankful the disaster happened a month before the breeding seabirds returned to their huge crowded colonies. More birds have been displaced by the Cardiff Bay Barrage Scheme although it is hoped some of the wildfowl and waders can be encouraged to winter on the Gwent levels. Rarities and marshland species can be sought at several sites in South Wales. Almost anything can turn up at Kenfig Pool - in 1997 Bittern, Penduline Tit and Purple Heron did so. Oxwich, an area of marshland on the Gower coast has one of the largest concentrations of Cetti's Warblers in the region and Bearded Tits have bred there in recent years.

Pembrokeshire has some of the finest stretches of coastline in Britain and a 120 mile coast path to enable the walker and naturalist to enjoy it all to the full.

The offshore islands, particularly Skokholm, Skomer and Grassholm, provide nesting sites for many thousands of seabirds, making it undoubtedly the most outstanding area for them in the whole of England and Wales. Gower, in West Glamorgan and on Swansea's 'doorstep', has equally magnificent cliffs and sandy bays. This district containing Oxwich, Worms Head and the south shore of the Burry Inlet, is good for a wide range of birds and holds the most southerly breeding Choughs in Britain.

The mild climate of south-west Wales is an important reason why several species are found in this region more commonly than elsewhere in Wales during the winter. Among these are included the Black Redstart, Firecrest, Bittern and waders such as the Greenshank, Ruff and Green Sandpiper. This part of Wales, jutting out towards the Atlantic, also yields many records of passing oceanic species such as the Sooty Shearwater and regular sightings of skuas and other migrating seabirds.

Locations of special interest

1. NEWPORT DISTRICT (Pembrokeshire)

Habitat – Various good habitats in the area include mature deciduous woodland (Gwaun Valley and Pengelli Forest), estuary of river Nevern, headland cliffs at Dinas and Preseli mountains.

Birds – Good variety of woodland species including Pied Flycatcher, Redstart and woodpeckers. On coast, seabirds, duck and waders, Chough and Peregrine.

Access – Leave A487 Cardigan to Fishguard road east of Felindre Farchog for Pengelli (Wildlife Trust West Wales), take Parrog sign in Newport for estuary. Turn right on A487 towards Fishguard for Dinas Head (signposted Bryn Henllan).

2. STRUMBLE HEAD

Habitat – Headland, cliffs and coast path.

Birds – This is far the best location in Wales for migrating seabirds.

Access – Unclassified road is signposted from Fishguard harbour. Follow directions through Goodwick. There is a hide for birdwatchers near the lighthouse.

3. ST DAVID'S HEAD AND DOWROG COMMON

Habitat – Coastal cliffs, gorse and lowland heath and bogs with pools.

Birds – Hen Harriers winter on Dowrog Common and Merlins and Short-eared Owls are sometimes observed. Whooper and Bewick's Swans winter. Water Rail and Grasshopper Warblers breed. Coast path at St. David's is good for migrating seabirds.

Access – Paths lead across Dowrog Common which lies on the A487 just out of St. David's. Much of the common is owned by the Wildlife Trust West Wales.

4. PEMBROKESHIRE ISLANDS:
SKOMER
Habitat – Largest island, close to mainland with steep cliffs and a range of other habitats.

Birds – The best bird island in southern Britain. Huge numbers of seabirds include Puffins, Storm Petrels and the world's largest colony of Manx Shearwaters, exceeding 200,000 pairs. Short-eared Owls usually breed. Rare migrants and accidentals often recorded.

Access – By boat daily from Martin Haven from April until October (except Mondays unless a bank holiday). Owned by the Wildlife Trust West Wales. Tel. 01437765462 for details.

SKOKHOLM
Habitat – Rather flat island further from the coast than Skomer.

Birds – Supports more Storm Petrels (over 6000 pairs) and fewer Manx Shearwaters in their subterranean burrows. Less cliff nesting birds but Skokholm has an even more impressive record for rare and accidental vagrants than Skomer.

Access – Day trips on Mondays only from Martin Haven. Full board for up to 15 visitors/apply W.T.W.W. for information.

GRASSHOLM
Habitat – Small rocky island 9 miles from the Pembrokeshire coast.

Birds – Immense colony of Gannets numbering 30,000 pairs, is the only one in Wales and the fourth largest in the British Isles.

Access – Visits can sometimes be arranged through the Dale Sailing Co. but restrictions may apply. RSPB reserve (landing not permitted).

RAMSEY
Habitat – Inhabited island close to St. David's.

Birds – Cliff nesting sea birds (no Puffins, few Shearwaters and possibly no Storm Petrels). Choughs, Peregrines and often Short-eared Owls breed.

Access – Boat trips round the island from St. Justinian. Owned by RSPB.

5. CLEDDAU
Habitat – Network of tidal rivers, muddy creeks, pools and woodland.

Birds – Large numbers of wildfowl and waders especially in winter. Little Egrets and Herons. Nightjars breed in woods near Oakwood.

Access – Many access points to tidal flats both north and south of the Cleddau. For Oakwoods turn left after Wathen travelling west on A40, then immediately right. For Wildlife Trust West Wales reserve at West Williamston (mainly salt flats), park at village.

6. THE GANN AND MARLOES MERE
Habitat – The Gann is a small sheltered estuary with fresh water pools. Marloes Mere is a flooded water meadow 3 miles distant.

Birds   –   The Gann records Egrets, wintering waders e.g. Greenshank and migrants such as Stints and Curlew Sandpipers. Marloes Mere has wildfowl, breeding Snipe, visiting raptors etc.

Access –   The Gann is about 1 mile from Dale on the B4327.For the Mere, turn first left in Marloes from Dale. Follow track to hide from head of car park.

7. TYWI / TAF / GWENDRAETH ESTUARIES AND CARMARTHEN BAY

Habitat –   Mud flats, tidal creeks, beaches, headlands and shallow bay.

Birds   –   Wintering waders such as Bar-tailed and Black-tailed Godwits and Sanderling. At sea, divers and sea duck, especially Scoters. Sea birds off shore. Snow Bunting, Spoonbill, Great Grey Shrike and other rare species are all possible.

Access –   Mostly via minor roads. Footpaths at Laugharne (Taf) and Ferryside (Tywy). Gwendraeth estuary can be reached from Salmon Point Scar and there is a car park at Cefn Sidan Sands via the Pembrey Forest Park.

8. WORMS HEAD, GOWER

Habitat –   Headland and miles of steep cliffs.

Birds   –   The most easterly nesting point for Kittiwakes, Guillemots and Razorbills along the South Wales coast. Shags, Fulmars, gulls and Choughs also breed.

Access –   From the car park at Rhossili. The head and 6 miles of cliffs is a nature reserve managed by various conservation organisations.

9. OXWICH

Habitat –   Dunes, marsh, extensive reed fen.

Birds   –   Cetti's and Reed Warblers breed, Bearded Tit has done so. Rare herons are possible, Bitterns are annual.

Access –   Turn left off A4118. From beach car park footpath runs along edge of dunes and marsh. Much of this scarce reedland habitat is managed by the Countryside Council for Wales.

10. PENCLACWYDD AND NORTH BURRY INLET

Habitat –   Freshwater pools and extensive salt marsh.

Birds   –   Wildfowl, waders, wintering flocks of Brent Geese. Sizeable flock of Black-tailed Godwit. To the west, the Inlet has divers and sea duck.

Access –   Turn left off A483 between Gorseinon and Llanelli. Penclacwydd is managed by the Wildfowl and Wetlands Trust. An admission charge is made but facilities are excellent.

11. SWANSEA BAY

Habitat –   Shoreline, mostly backed by urban development.

Birds   –   Scarce gulls, notably the Ring-billed Gull.

Access –   Blackpill, the most important site lies along the A4067 just east of the B4436.

12. KENFIG POOL
Habitat – Sand dunes, heath, lagoon and lakeside reed beds.
Birds – One of the best sites in Wales for scarce migrants and visitors. Water Rail, Reed Warblers and sometimes Yellow Wagtail breed. Nearby Eglwys Nunydd Reservoir has wintering wildfowl, including (sometimes) Smew.
Access – Turn off junction 37 of M4 towards Porthcawl, turn right through village of South Cornelly. Owned by Glamorgan County Council.

13. TAL-Y-BONT RESERVOIR
Habitat – Fresh water lake fed by small river.
Birds – Wintering wildfowl at shallow south end of reservoir. Great Crested Grebes breed.
Access – Unclassified road running through Brecon Beacons from Talybont-on-Usk.

14. WYE VALLEY (GWENT)
Habitat – Extensive and sheltered deciduous and conifer woodlands.
Birds – Best area in Wales for Hawfinches. Also Nightjars and Woodcock.
Access – Many woodland footpaths open to the public.

15. WENTWOOD FOREST
Habitat – Conifer and deciduous woodland.
Birds – Nightjar and Crossbill breed. Formerly Firecrests did so.
Access – Take unclassified roads to the left of A48 near Parc Seymour or Llanfaches.

16. LLANDEGFEDD RESERVOIR
Habitat – Large freshwater lake with recreational access to the public.
Birds – Wintering water birds. Migrants such as Black Terns. Great Crested Grebes breed.
Access – Minor roads from Usk or Pontypool.

17. SEVERN ESTUARY / GWENT LEVELS
Habitat – Large, strongly tidal estuary and drained water meadows.
Birds – Wildfowl and waders on estuary especially in winter. Water Rail, Yellow Wagtail, Snipe and Tree Sparrows breed on the levels. Birds of prey in winter.
Access – West of Newport, Peterstone Wentlooge is a good focal point, while to the east on the Caldicot Levels the shoreline is accessible in two or three places. The Gwent Wildlife Trust owns the Magor Reserve which has pools and scrub (permit required).

18. CARDIFF
Habitat – Many good bird habitats near or in the capital include the city parks, Cosmeston Pool and Llanishen Reservoir. Also the Taff and Ely rivers and the Rhymney estuary.
Birds – **Parks:** Sparrowhawk, Heron, Green Woodpecker etc.

24

Lakes: Wintering wildfowl, Smew, sometimes Bittern. Black Terns on migration. Breeding Pochard, Tufted Duck and Great Crested Grebes.
Rivers: Dipper and Grey Wagtail, Heron.
Estuaries: Wildfowl and waders.

Access – Generally free with ready access to the public. At Llanishen there is a membership fee for admission to the two lakes.

19. LAVERNOCK POINT

Habitat – Headland of limestone cliffs.
Birds – Passing seabirds may include scarce shearwaters and skuas, (especially the Pomarine Skua) and Storm or Leach's Petrels.
Access – Take an unclassified road off the B4267 Penarth to Barry.

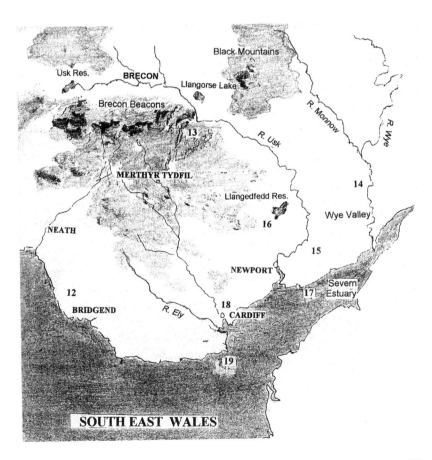

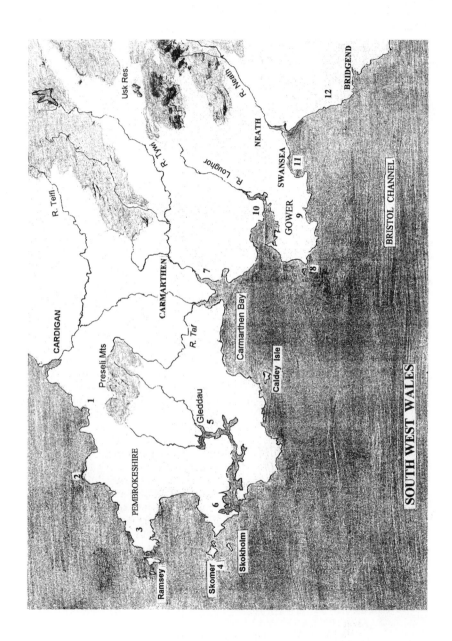

Ramsey
Skomer
Skokholm

PEMBROKESHIRE

CARDIGAN

R. Teifi

Preseli Mts

R. Tywi

Usk Res.

R. Neath

NEATH

BRIDGEND

R. Loughor

SWANSEA

GOWER

Gleddau

R. Taf

Carmarthen Bay

Caldey Isle

CARMARTHEN

BRISTOL CHANNEL

SOUTH WEST WALES

1
2
3
4
5
6
7
8
9
10
11
12

CHAPTER ONE
FARM AND WOODLAND

Wales is predominantly a country of rolling green hills cut by deep valleys, whose slopes are clothed in deciduous woods of oak and beech. Sheep graze the upland pastures, while cattle browse on the richer meadows in sheltered vales. Where drainage is poor, patches of reed cover unimproved grassland, while bracken slopes reach upward towards the open moor. Clumps of alder and birch gather around gurgling streams, and thickets of bramble or blackthorn occupy neglected corners of fields. In the poorer homesteads, crumbling stone walls and half derelict barns surround old grey farmhouses where the occupants work long hours to extract a living from the land. Elsewhere, pristine white-painted dwellings are surrounded by vast modern barns which speak of better fortune. Nineteenth century chapels, still attended by the vanishing faithful of a declining rural population, stand defiantly in remote places, surrounded by lichen covered tombstones, ivied walls and sheltering groves of beech or pine.

This Welsh rural scene is still commonplace in many parts of Wales. Woods and farms are inextricably linked in this beautiful but sometimes austere setting, so we have taken them together in this chapter. There are some special birds too, to be seen in the woods and in the skies over this typical Welsh countryside. Buzzards mew their high pitched calls from the depths of their nesting woods or as they circle on broad, slightly upturned wings high above the hills. Here they take advantage of currents of rising warm air, which gives added lift to their robust frames. They seem to revel in the warmth and radiance of bright sunshine. Sometimes four, six or even more birds may be seen soaring together, especially after hours of incessant rain when the birds have been forced to huddle inactively in the shelter of the woods. A pair of Crows will fly up to launch an attack on any passing Buzzard which dares to stray too close to their nest, harrassing the big brown mottled bird mercilessly until it is out of range and their young fledglings are safe from its attentions.

There may be other large birds circling over the trees or soaring high above the valley. With each passing spring, young pairs of Kites display over new nesting woods in valleys where breeding cannot be recalled by anyone alive today. In districts

Pied flycatcher at nest

well beyond their former tenuous stronghold in Mid Wales, where a few pairs clung to survival for half a century, Kites may once again be seen rising in the same thermals, or hunting rabbits on the same sheep walks as the Buzzard. The Kite seldom seems to land, its slim frame propelled easily on wings spanning five feet and its direction steered by delicate rudder movements of its long forked tail. Ravens, not to be outdone by their more illustrious companions, circle round like large black hawks, and tumble towards the ground and chase each other over the woods in early spring frolics. Raven and Buzzard are widespread and familiar birds of the Welsh countryside and if current progress continues, we may soon be able to say the same of the Kite. In springtime, Sparrowhawks may be seen displaying above the valley woods, or sometimes even the large and powerful Goshawk. Both have the same distinctive silhouette of broad relatively short wings and long tail, but the Sparrowhawk is much slimmer and smaller than the stoutly built Goshawk.

In the woods, especially near streams and areas with decaying trees for nesting, you can see other characteristic Welsh birds. Outstanding among these is the lovely Pied Flycatcher. Listen for the song consisting of just four or five notes. The pied male bird is quite superb. His neat appearance is glossy black above, with white forehead and wing bar and matching white underparts. The female is similar but is brown where the male is black. In some woodlands in Mid Wales the Pied Flycatcher is almost the dominant species, as common as Chaffinch or Blue Tit. It readily takes to using nest boxes and this has helped it to expand its range eastwards, further into the English counties than ever before. Nevertheless, the sessile oak woods of Wales support higher densities of this species than anywhere else in Britain.

The Spotted Flycatcher looks quite different. It has a few throat spots and purplish-brown wing primaries but its undistinguished brown and whitish

Wood Warbler

28

plumage easily goes unnoticed and the bird may often be overlooked. This is a pity because it has some interesting habits. If the bird you are watching flies from its perch, dashes in pursuit of an airborne insect and returns repeatedly to the same perch, it is almost certainly a Spotted Flycatcher. Like the Pied Flycatcher it is fond of places close to water but quite often it will make its neatly woven nest on a barn ledge or against a shed in a rural garden. This species usually arrives in late May from winter quarters overseas and probably June and July are the best months to see it.

The Redstart nests in holes in trees or crevices in stone walls or buildings. In summer it is common in open woodlands and hillsides where there is a good scattering of mature trees and near farms in well timbered places. Much of Wales is ideal territory and like the Pied Flycatcher it is found in high concentrations here. The Redstart is one of our most colourful birds. The male has a black face, pale grey matching crown and red-orange breast. Both sexes display a flash of red tail as they disappear into the canopy of a tree or dive over the top of an old stone wall.

In woodlands of beech and beneath other trees where the close canopy prevents light reaching the ground, the accelerating trill of the Wood Warbler is commonly heard in springtime. Here there is no understory of shrubs and brambles and a carpet of leaves and grass suits the Wood Warbler well. It is greenish above and pale below with a yellowish throat and is closely related to the more sombre Chiffchaffs and Willow Warblers which are described later in the chapter. After feeding, the incubating bird can be heard calling its melancholic 'pee' note, which persists until it drops quietly into its domed leafy nest, hidden among the carpet of dead leaves on some wooded slope.

So far we have concentrated on some of the 'special' birds of Welsh farms and woodlands, but the fact is that most of the thirty or so commonest species in Britain most frequently seen on a country walk can be replicated with minor variations whether you are in Kent, Cumberland or Carmarthenshire. Many of these are almost too familiar to need description: the Robin, Blackbird, Song Thrush, Wren, Blue Tit, Great Tit and Chaffinch are among our best loved birds. The Chaffinch is probably the most abundant species in Wales and is most easily recognised by its dull-pink breast and white wing and shoulder patches. All of these birds are to be seen in country lanes and gardens everywhere. Unfortunately the ever popular Song Thrush has declined rapidly in recent years (fortunately though it is thriving well in the new conifer forests) whereas the number of Robins and Blackbirds have held up reasonably well. The Thrush has a more meaty diet and is particularly fond of snails, which leads one to wonder how much the Thrush's demise has been caused by slug pellets or insecticides. One also wonders what effect the ravages of Magpies, inveterate robbers of both eggs and nestlings, has had on the population of this and other garden birds. The handsome Magpie is omnipresent even in urban gardens, its numbers inflated by the plentiful supply of carrion from rabbits, hedgehogs and other animals killed by the ever increasing volume and speed of traffic on our roads.

At this point we will introduce the birds of the countryside by taking you on a walk in springtime starting from a village garden, and travelling through typical country lanes to see what birds are likely to be encountered as the habi-

Dunnock at Nest

tat changes. If you have a bird table or feeder you will be able to enjoy the antics of Great Tits and Blue Tits at close quarters. Depending where you live and the kind of food you place for them, other feathered visitors will arrive at the table. Coal Tits and Siskins will take the peanuts, Chaffinches the seeds, and House Sparrows and Greenfinches, both seeds and nuts. In west Wales the Tree Sparrow is very scarce, but a pair turned up regularly to feed all winter at a friend's garden bird table near Aberystwyth. Closer to town, a Blackcap and a Chiffchaff, which had apparently 'forgotten' to join their peers in the southward migration, were almost daily guests in another garden. Ringing records in fact, show that wintering Blackcaps are invariably immigrants from central Europe. One December morning, I awoke to find a male bird busily demolishing the clusters of yellow flowers on the mahonia bush in my own garden and this bird stayed nearby for the next four weeks.

Robins and Hedge Sparrows habitually rummage on the ground, feeding on pieces of soaked bread or whatever suits their digestion. That the 'Hedge Sparrow' or Dunnock is not really a sparrow (it is an accentor), can be deduced by its un-sparrow-like thin beak. It is a modest, quiet bird, in both voice and plumage, unlike the Wren which bursts forth with vigorous song at every opportunity. Both species skulk in woodland undergrowth and brambles, the Dunnock in mouse-like fashion, the Wren with loud scolding interjections. The Dunnock lays the most beautiful turquoise blue eggs in its hedgerow nest, the Wren wedges its domed structure in any crevice in which it will fit. A pair placed their nest in the saddle of an old bicycle in my garden shed which must rank as a rare example of a mobile home in the bird world! In contrast to these little sober coloured birds, two of the most striking species which you seldom have the opportunity to observe at close range, the Nuthatch and Great Spotted Woodpecker, quite frequently grace garden bird tables. Whatever birds turn up, identification will be made easier when you can enjoy them from your kitchen window.

Greenfinches are most common among shrubberies and conifers, so they will be readily seen in large gardens and villages. They usually draw attention with asthmatic wheezes delivered from a perch, or loud trills as they bound from one tree to the next. The females often look rather sparrow-like but the yellow edge to the wing is always a safe distinction. The cinnamon-tinted grey Collared Dove is also usually found here too. Its coo-coooh-coo calls uttered from the branch of a tree are to be heard in most villages now, although the species only began nesting in Britain in 1955. Starlings will be seen flying like darts among the houses or rattling their curious throaty notes from a television aerial or chimney.

As you leave the towns and villages and stroll along country lanes, members of the crow family are the most obvious birds to be seen on the open fields. The black corvids will usually be Carrion Crows, Rooks or Jackdaws. Usually a pair or two of the conspicuous Magpie will be heard chattering or squabbling within the canopy of hawthorn bushes at the edge of the field. As you approach an old stone farmhouse surrounded by barns and loosely constructed stone walls, the Pied Wagtail, flicking its tail up and down and flitting about the farmyard in search of insects, is likely to be one of the first birds to come to your notice. After about mid April there should also be Swallows which make their half cupped nests on the barn rafters, and maybe House Martins, dashing and twisting in the air in their unrelenting pursuit of airborne insects. Note the white rump of the inky blue-backed House Martin which has a forked tail, but lacks the long streamers of the Swallow. Goldfinches like to feed on thistles, burdock and other wild seeds on rough patches near the farm buildings. They are often the second commonest finch after the Chaffinch over much of rural Wales. Look for the broad yellow wing patches when the bird is flying or at rest, and listen to the tinkling medley of notes as it sings from a telegraph wire. The Goldfinch is a particularly attractive bird with a maroon-red face and golden wing bars.

On reaching a spinney of trees where there is an understory of light sufficient for the growth of brambles and shrubs, more species can be confidently expected. A Jay may screech from a thicket of trees and fly in its laboured fashion to a tall holly at the far corner of a field. In good light the beautiful Jay is unmistakable. At other times, in poor light in overcast weather conditions, its colours can be totally subdued. The dusky pink plumage, and black and white wing primaries decorated with a patch of blue and white stripes, can surprisingly appear a uniform sombre grey. Look for the white rump which is noticeable as the bird flies away from you. A loud clap as wings become airborne will turn your attention to another medium-sized bird, the Wood Pigeon. The Wood Pigeon's flight is fast and erratic compared to the weak but direct flight of the Jay, and its habit of swooping upwards before gliding down on half folded wings to land in a tree is totally characteristic of its family. Its larger size, white neck and wing patches, distinguish this species from the Stock Dove which has a black border on the rear edge of its wings.

Any medium sized bird which is not black, nor white (you may well see Black-headed or Herring Gulls over the open country) is likely to be one of just a few species. Pigeon, Jay or Mistle Thrush are the most likely. In winter, thrushes in flocks are more likely to be Fieldfare or Redwings than Song or Mistle Thrushes. The grey-brown Mistle Thrush, like its close relatives, has a rather springing flight. Its loud ticking alarm or call notes are quite unique and easily recalled once the sound has been learned. The similar sized male Sparrowhawk may dash across your path or circle above the wood or nearby fields, its penetrating yellow eyes seeking out its next quarry which is likely to be a small bird.

The spinney or woodland edge we have described will nearly always harbour several species of warbler, which are usually frustratingly difficult to see. Just as you have a bird focused in your binoculars, it moves behind a thick branch or into the cover of foliage. If you can recognise the song, identification will be made much easier; that of the greenish-brown Chiffchaff is not at all difficult - it simply repeats chiff chaff - chiff chaff quite incessantly. The Willow Warbler looks almost identical but sings a sweet 'catchy' melody which is one of the commonest of all bird songs in spring and summer. Both species make domed nests on or near the ground but the Willow Warbler chooses a rather open position cleverly concealed in patches of grass or bracken, while the Chiffchaff usually builds its wren-like nest in low brambles or other impenetrable plants.

Long-tailed Tits

Garden Warbler and Blackcap are easier to tell apart providing you are able to see them properly - and that isn't always easy among the canopy of trees, where leaves, stems and dappled light obscure the view of birds which often seem to play a deliberate game of 'hide and seek'. The greyish male Blackcap has a black cap and the female a brown one. Both species sing intricate and beautiful warbling songs which can only be separated after some effort and experience. The Blackcap's is shorter and more melodious, while that of the Garden Warbler is sustained longer and is more scratchy. The Garden Warbler is the 'proverbial brown warbler' and has no clear-cut distinguishing features. In many parts of Wales it is the commoner bird. The nesting territories of the two species overlap considerably in the same woods but as a rule Blackcaps prefer lower, more sheltered situations, while the Garden Warbler is common in upland woodlands and plantations. Both species build fragile nests of rootlets and hair on a thin horizontal branch of a bramble or small bush at the edge of a wood.

Mature woodland, or damp low lying woods whose trees have become riddled with cavities due to age and decay, attract hole nesting species. Marsh and Willow Tits are neat, small, brownish tits with black caps, white cheeks and fine black bibs. These two species are very hard to tell apart, although you will have little trouble distinguishing them from the Coal Tit, which has heavier head markings and a diagnostic white patch on the nape. Both Marsh and Willow Tit nest in holes in rotten trees but the 'Willow' excavates its own cav-

Great Spotted Woodpecker

Redstart (male)

Chough

Fulmar

Barn Owl

Whinchat (male)

Golden Plover

Ring Ouzel

ity. It has a larger head than the Marsh Tit and a faint, pale wing patch. Its scolding 'char' note is deeper and harsher than that of the other bird. This unique nasal buzzing note is often the first indication you have of the presence of either species. Despite its name, the Marsh Tit has less predilection for damp sites than the Willow Tit and may be found in relatively dry woodland.

These two tits are not uncommon but I would not guarantee to see either of them during a single day's bird watching in my district. You will however, almost certainly see Long-tailed Tits, especially later in the year when they move about in family parties or in flocks during autumn. They are birds of the hedgerows and scrubby corners of fields rather than the deep wood. I must admit they are among my favourite birds - tiny pinkish-brown and white tits with long tails, they move acrobatically in search of seeds or insects among the twigs of birch, alder and other trees. They make arguably the most wonderful nest of all British birds - an intricate ball of lichens, mosses and spiders' webs lined with hundreds of feathers. The small entrance is near the top of the nest, which is often placed in a gorse or thorn bush, and contains up to eight or nine eggs. Nests can be ridiculously easy to see, and the birds so readily give their nest sites away early in the season, when predation is heavy.

The Nuthatch is one of our most characteristic species in Wales. It has a wide repertoire of loud calls which it utters from the trees in springtime. When observed at close range, the blue-grey back, peach coloured underparts and dark line through the eye are striking. Unlike the Woodpeckers, it is able to walk down the trunks of trees as well as up them. If its nesting hole is too large, it bricks it up with mud which hardens to the correct size, just large enough for the adults to leave and enter the nest hole, and small enough to keep out would-be predators. The Treecreeper is a small brown bird with white underparts and a long curved beak, which it uses to prise insects and grubs from the cracks in the bark of trees. It is a delightful little bird which will spiral jerkily around the trunk of a tree, moving upwards until it decides to fly down to the base of the next, when begins the process all over again. Treecreepers usually nest behind dead bark. They can be detected sometimes by their thin 'si si si' calls which are not unlike those of some of the tits.

The Woodpeckers are striking birds having colourful and strongly marked plumage. In dull light they can readily be picked out by their dumpy appearance and strongly undulating flight. On landing, more often than not they cling in characteristic vertical fashion against the trunk or branch of a tree. The two large woodpeckers, the Great Spotted and Green Woodpeckers, have very different plumage. There are other differences too. If you hear sharp drumming sounds it is one of the Spotted Woodpeckers (we will refer to the Lesser later). The Green Woodpecker does not drum. It does, however, make a loud though not very mirthful laughing call which immediately arrests your attention. Despite its red crown and green back and wings, in diffuse light the colours can appear surprisingly subdued and it may occasionally be mistaken for the similar sized Mistle Thrush. When seen from the rear as it flies away, the vivid yellow rump is a noticeable feature of this woodpecker. The boldly marked, pied Great Spotted Woodpecker has a distinctive sharp and piercing 'tuck'note, which is easily recognised again once the sound has been associated with this bird.

The Great Spotted Woodpecker is likely to bore its nest hole in a rotten birch or some other decaying tree in the deeper part of the wood whereas the Green Woodpecker has a liking for more open ground. The latter favours parkland with old trees and heather or bracken-clad heaths, where it will often feed on the ground. If it finds an ants' nest it can quickly devour the insects, licking them up on its long sticky tongue. Whether you can expect to see a Green Woodpecker regularly much depends upon which part of Wales you live in. In Ceredigion I have seen and heard this species infrequently, and there are believed to be only about five pairs now remaining in the whole of the county. During a recent visit to Cardiff, I saw Green Woodpeckers twice in city parks on the same morning, as many

Green Woodpecker

as I would expect to see in a whole year near to home. At a large cemetery in the city, a young bird was perched on a tombstone, hungrily awaiting grubs fed to it by its two industrious parents. This species is far commoner in south-east Wales, from Carmarthenshire eastwards through Breconshire, Glamorgan and Gwent, than it is elsewhere.

The Lesser Spotted Woodpecker is an elusive species. One reason for this is its small size, barely much bigger than a sparrow. It drums more quietly than the 'Great Spotted', but on average for twice as long. It seems most often to frequent damp woodlands close to water. This woodpecker is scarce and local in Wales but may be discovered by its piercing 'quee' calls repeated in quick succession.

Like the woodpeckers, owls are a fascinating family of birds, although being largely nocturnal they less often figure among species listed on a daytime birding excursion. By far the commonest is the Tawny Owl, most often detected by its well known quivering hoot delivered at night. This species is naturally a denizen of deep woodlands although it will nest in cavities in hedgerow trees. Sometimes you will hear one during the daytime or disturb this bulky brown bird from its roosting tree, when it will lumber off through the trees to find a quiet place to resume its slumber.

The Little Owl prefers open farmland and is scarce in Wales, yet it is plentiful in all of the English border counties. There are some resident on the Gower and in Powys and more occur along parts of the coast of north-east Wales. I have only seen it three times in Mid Wales, twice near Welshpool and once at Capel Bangor, Aberystwyth. Each time it was huddled up like a grey ball, asleep on a telegraph wire. It is more diurnal in habits than the Tawny Owl and has a liking for riversides where there are pollarded willow trees, a habitat not at all prevalent in Wales.

Much concern has been expressed about the plight of many peoples' favourite, the Barn Owl, over the past decade or so. Its decline may be related to food supply, competition with the more powerful Tawny Owl, disease, or the disappearance of traditional barn nest sites. Many old stone buildings with lofts have been replaced by modern barns but some farmers are fitting specially adapted nest boxes to their new barns to accommodate this unique bird. Some nest however, among hay or straw bales in quite new sheds! The main problem is the decline of rough grassland and with it, the supply of small mammals upon which the owl feeds. In Wales, Barn Owls have fared better than on the arable farms of eastern England, where the decline has been far more serious.

Looking quite eerie at dusk in the fading twilight, this white and sandy owl flies delicately on silent wings along ditches, field edges and similar places where there is rough ground. Searching for field voles, rats and other rodents, it will suddenly pounce, diving to one side, piercing its unfortunate prey with razor sharp talons. It returns to its nesting site, often through a wide window entrance in the disused loft of an old barn near to the farmhouse, to feed its three or four fluffy white young. Sometimes it chooses a nest site in a hole in a mature tree.

In early July people in my local village informed me that a strange bird, believed to be a Nightjar, was calling almost every night close to the village.

There was no heath in the vicinity suitable for a Nightjar, and I was intrigued to establish the true identity of this bird. When I first heard it, the strange 'querk, querk,' rasping calls, which sounded like something between a hiss and a snore, continued non-stop for over half an hour. The performance continued regularly on most nights for the next two weeks, and I never managed to get a glimpse of the bird, although one or two had seen a whitish bird near the church. Perhaps this owl was an unmated male whose forlorn romantic appeals went unanswered!

We have noted that Redstarts and Pied Flycatchers, among others, are found in greater concentrations in Wales than in most parts of England, where many areas in the lowlands have none of these birds at all. Some species however, especially those finches and warblers which prefer dense lush vegetation, are scarcer in many parts of Wales than in England. Yellowham-

Little Owl

mers, Bullfinches and Lesser Whitethroat number among such species. Here in Wales, commons and overgrown hedgerows are good places to find them. Sometimes disused railway lines with their tall bushes and impenetrable tangled undergrowth, the result of years of neglect, become highly suitable.

Last year we located a pair of Lesser Whitethroat by exploring just such a habitat. The Lesser Whitethroat, typically a species of tall hedgerows and thickets of eastern lowland England, has been increasing in Wales for some years. We detected our bird by first identifying and then following the source of the song, a rather unmusical series of monotone notes rather like those of the Chaffinch, but without the final flourish. Some typical scratchy Whitethroat notes were also uttered from the thick tangle of thorn bushes, as if to confirm this secretive bird's identity, in case we might have been inclined to pass this warbler off as some other species. The male has a distinctive dark patch round the eye and the back and wings are more olive and darker than the lighter brown of the Common Whitethroat. We waited patiently for several minutes and were rewarded with a view of one of the neatest and most delightful warblers I can remember seeing for a long time. First the male appeared and then his mate emerged from the cover of dense blackthorn scrub. For a few minutes they chased each other around the bushes, then dived into cover and were gone without further trace. We stood gazing at the spot where they had disappeared, wondering whether we had really seen them at all!

There are other elusive and scarce birds to be looked for. The Hawfinch is only frequent in isolated, widely separated pockets in Britain. Kent, the home counties, parts of northern Lancashire, the East Midlands, and the area close to the Wye Valley and Forest of Dean are among the most favoured areas for this species. Gwent therefore offers the best chance of seeing a Hawfinch in Wales, although pairs may be met with elsewhere, especially in eastern districts. The Hawfinch is a large finch with a disproportionately large head and stout beak and it flies in a heavy bounding fashion. It is however, a shy bird which feeds high in the canopy of trees like oak, hornbeam and wild cherry. In late winter and early spring it is active, uttering loud 'tick-tick' calls in its forest haunts and this time of the year is probably your best opportunity to see one. If you are very lucky, one may turn up at your bird table in winter!

A bird more often heard than seen is the plump Quail, the diminutive partridge of Biblical fame. A friend phoned me to say she had heard a strange bird in a clover field not far from Aberystwyth. It had a distinctive trisyllabic call 'quick qui quick', which was constantly repeated as it moved around the field, in total concealment amongst the crop. The Quail is heard in many districts in some years, whilst in others it is almost completely absent. If we experience warmer summers, the call of the Quail could become a more familiar sound. Ironically the larger partridges, Grey and Red-legged, are scarcely more likely to be encountered in most parts of Wales than the Quail. The best places to see them are on Anglesey or places close to the eastern perimeter of the country. Only the Pheasant among this group of game birds manages to thrive in parts of Wales, finding protective cover for feeding and breeding in the woods, its numbers reinforced by regular releases of pen-reared birds.

There is much unspoiled countryside in Wales but even here some species, especially those which nest on the ground, in the open field, or low down in hedgerow vegetation, are finding survival hard going. On 'improved farmland' the grass is regularly re-seeded, fertilised and often heavily grazed by sheep, creating what is commonly called the 'billiard table effect'. The place where the food is actually produced, namely on the fields, is poorest in wild life. Not so many years ago, there was greater diversity in farming and a variety of arable crops were sown. There were ponds and patches of herbage and wild flowers at the edge of fields. There were wild fruits and seeds, insects and stubble which supported a healthy population of birds. With modern machinery it is now so easy to cut down trees, uproot hedges, and drain wet meadows. Crops are sown, rolled, fertilised, sprayed with insecticides, and harvested, in a shorter space of time than was ever possible before. At the same time there has been an increase in mammalian predators which find it easier to locate ground nesting birds in an environment which provides less protective cover. The result is that Lapwings are a thing of the past on nearly every farm in the country while Skylarks are consigned to the moors and roughest pastures. Corncrakes and most Corn Buntings disappeared a long while ago and Yellowhammers have declined along with the fieldside herbage in which they like to build their nests. Wet pastures with reeds, waste ground and common land still support Meadow Pipits and Skylarks and still provide nesting sites for a declining population of Curlew, Snipe and Whinchats. Kestrels searching for mice and shrews still hover overhead, waiting for the slightest movement to betray the presence of a small rodent in the long grass. Tree Sparrows and Stock Doves are to be found in some areas, especially where there is some mixed farming, notably in parts of south and east Wales.

In the past 50 years some species have ceased to breed in Wales. We have already mentioned the Corncrake, which could not cope with the high level of mechanisation on farmland where it used to nest in the hay meadows and in corn. The Cirl Bunting used to be found in coastal districts in Wales. Now, in Britain it is more or less confined to South Devon, where about 120 pairs have increased to 400 pairs as a result of co-operation between farmers and conservation bodies. It was discovered that the buntings were starving in the winter as a result of the modern practice of removing all the stubble from the arable fields. The Red-Backed Shrike last bred in Wales in about 1950 and has now finally ceased nesting in its last remaining stronghold in East Anglia.

Hawfinch

The Butcher Bird as this shrike is sometimes called, occasioned by its habit of impaling frogs, mice and other prey on a thorn tree larder, used to be found on shrubby commons. The reason for the slow withdrawal eastwards and final disappearance of this species from Britain as a breeding species remains a mystery. It still turns up occasionally in Wales on autumn migration. The Wood-lark ceased to breed on Welsh farmland after the cold winter in 1963. Only three years before, the great naturalist William Condry heard 12 singing birds within three miles during a cycle ride not far from Aberystwyth. Perhaps the Woodlark will return again to pour out its liquid fluting song as it circles high above the gorse and bracken slopes of South and Mid Wales. This species still occurs occasionally in spring or autumn, usually in South Wales.

We must all be concerned about the plight of some of our farmland birds but perhaps with decisive government action the downward spiral could be halted. The autumn sowing of grain, for example, means that there is no win-ter stubble for finches, buntings and partridges and a crop too tall in spring for Larks or Lapwings to nest in. Why not in selected cases, subsidise spring sow-ing and in others, pay the farmer to leave alone or create wet pasture, flower-ing hedgerows and 'unimproved' patches of ground which are so important to wildlife? New cultivars of perennial rye grass began to be developed over 40 years ago. Rye grass is tough, grows rapidly, and produces good fodder. Unfortunately it also effectively stifles other flowering and seed producing plants. Furthermore, silage is ready for cutting in May before ground nesting birds can rear their young. Could not a percentage of farms be encouraged to use other species of grass? The issues are obviously complex and require a close working partnership between farmers, conservation agencies and gov-ernment. Perhaps food production has simply become too efficient and too competitive for the good of wildlife and a new balance is needed! Sometimes though, finding a specific cause of a species' decline and taking effective ac-tion, as in the case of the Cirl Bunting in Devon, can lead to encouraging re-sults. In the Hebrides, as a further example, the number of breeding Corncrakes has markedly increased following a change in the practice of harvesting hay. By reaping the crop starting from the middle of the fields it has been found that far fewer fledgling birds fall foul of the plough, since their parents are more easily able to lead them to the edge of the field and safety.

Leaving the issues of conservation aside, we will finish the chapter on a lighter note. There is just a chance that you could come across something re-ally rare and exotic in rural Wales, especially in springtime. In 1995 a pair of magnificent Hoopoes bred on a farm in Montgomeryshire and in 1993 a Bee-eater was seen on a telegraph pole in Ceredigion by an astonished birdwatcher. The equally exotic Golden Oriole is discovered rather more often. Last year one was heard singing in a garden in Aberystwyth. It was discovered by a postman who heard the fluting 'oriole' notes emanating from a tree in his mother's garden and reported it to a local birdwatcher. We may hear more such stories if Welsh rain is replaced by drier summers!

FARM AND WOODLAND - species to look for:

A)

Chaffinch	Robin	Jackdaw	Meadow Pipit
Blue Tit	Wren	House Sparrow	Swallow
Great Tit	Carrion Crow	Starling	Herring Gull
Blackbird	Magpie	Wood Pigeon	Black-headed Gull
Dunnock	Rook	Pied Wagtail	Mallard

B)

Buzzard	Song Thrush	Sand Martin	Chiffchaff
Pheasant	Swift	House Martin	Garden Warbler
Collared Dove	Skylark	Redstart	Blackcap
Mistle Thrush	Jay	Coal Tit	Greenfinch
Fieldfare (winter)	Raven	Long-tailed Tit	Goldfinch
Redwing (winter)		Willow Warbler	Linnet

C)

Heron	Stock Dove	Spotted Flycatcher	Treecreeper
Kestrel	Green Woodpecker	Pied Flycatcher	Wood Warbler
Sparrowhawk	Great-spotted Woodpecker	Willow Tit	Yellowhammer
Lapwing	Cuckoo	Marsh Tit	Bullfinch
Tawny Owl	Nuthatch		

SCARCE SPECIES

Barn Owl	Lesser Spotted Woodpecker
Tree Sparrow	Kite
Lesser Whitethroat	Hobby
Red-legged Partridge	Hawfinch
Grey Partridge	Quail
Little Owl	

Goldfinch

Notes: (1) Generally, birds in group (a) are more often seen than those in group (b), and those in
(b) more often than those in (c). Some, such as the warblers and flycatchers, are summer migrants, most are resident. (see check list or text for details).
(2) Heron, Black-headed Gull, Herring Gull and Mallard are more familiar in waterside habitats but all are frequently observed on farmland.
(3) Many of these species, especially the commoner ones, occur widely in other habitats.

CHAPTER TWO
THE COAST IN SUMMER

A birdwatching outing or holiday by the sea offers the prospect of a wide range of species and as an additional reward, the opportunity to enjoy some of the outstanding and varied seascapes of the Welsh coast. If we start mid-way between north and south from a good vantage point along Cardigan Bay, the shoreline sweeps into the distance in both directions, offering views on clear days which span nearly the whole length of Wales. To the south lies New Quay Head, Cardigan Island and beyond that the bleary purple outline of the distant coast of Pembrokeshire. To the north the long arm of the Lleyn Peninsula stretches out into the Irish Sea, its steep hills of igneous rock thrusting upwards like islands, while the lower ground remains invisible below the sea on the distant horizon. Bardsey lies a few miles to the west, a bird-rich island well known for its regular migrant visitors and the accidental vagrants which turn up on its shores.

In the south-west, a coast path winds round the entire seaboard of Pembrokeshire providing an ever changing scene of sheer rock faces, coves and sweeping bays of white sand. Off-shore islands swarm with birds and coastal woods of sessile oak creep down almost to the shore. Bold headlands thrust to meet the inrushing Atlantic swell and temper the force of south-westerly gales which so often assault this part of the coast. The limestone cliffs and beaches of Gower, separated from Pembrokeshire by the low-lying districts of south Carmarthen, provide recreation for the residents of Swansea and visitors to the area. Northwards along the west coast, the same pattern of cliffs, interrupted by bays and snug coves, continues far into Ceredigion.

Walking along steep undulating tracks can be tiring as well as exhilarating. Tortuous trails hug the uneven coastline and climb steeply to the crest of successive hills, before descending to sea level again. Here the path may cross a wooden bridge beneath which a cool stream spills over the beach, where its fresh waters mingle with ocean brine. In small villages or hamlets, the weary hiker can rest and seek sustenance at the pubs and restaurants interspersed along the way.

On bright blue mornings when strong sunshine has evaporated the early dew, the air is filled with the fragrance of bluebells and the coconut-tang of blossoming gorse. This ubiquitous plant flowers throughout the year but is especially resplendent in springtime. Primroses, celandines, orchids and a wealth of wild flowers line the pathways or steep slopes, while white saxifrage and pink thrift cling to the rocks, providing a plethora of colour. The heady fragrance and vivid colour of flowers, the ever-changing shades of ocean blue and green and the rhythmic sound of the ebb and flow of tide, are a natural therapy and provide an excellent remedy for the stress and hassle of daily life.

In North Wales the scenes are just as varied and delightful. The Lleyn Peninsula is for the most part treeless with miles of wild exposed cliffs to the north and west, where bays are tucked away in quiet corners of the coastline. The sheltered southern edge of Lleyn is less remote with fine beaches and tranquil off-shore moorings for sailing boats which bring holiday makers to the resorts of Abersoch, Criccieth and Pwllheli. The north coast of Wales further east is also popular with visitors, who still flock, though in decreasing numbers, to the towns of Rhyl and Colwyn Bay in summer. Llandudno lies in the lee of the Great Orme, whose high limestone cliffs provide one of the best refuges for nesting seabirds on the northern mainland. Here you can watch Shags and Guillemots plunging in the sea, while Kittiwakes and Fulmars circle round below the sheer cliffs which lie to the north and west of the town.

Not all of the Welsh coast is lined by rugged cliffs interspersed with bays. There are long stretches along the North Wales coast and in the west between Harlech and Aberaeron where the shore is far more accessible, consisting mainly of sand and shingle lined beaches. Some of these are fronted by caravan parks, where families share the fringes of pebbly beaches with Ringed Plovers and Oystercatchers which still manage to find places to raise their young against human competition for space.

The coast of Wales is intersected by many estuaries and these provide important focal points for birdwatching. The tidal flats and marshes of the Severn, Cleddau, Conway and many others, are all good for birds. Even better are those of Burry Inlet on Carmarthen Bay and the Dee, an estuary of international importance for wintering waders such as Knots, godwits and Grey Plover. In these habitats large areas of nutrient rich sand are exposed at low tide. Mud flats and brackish marsh higher along the estuaries provide rich pickings for large numbers of wildfowl and waders, particularly in wintertime, when vast numbers of birds arrive from the north. In spring and summer, there are smaller yet still significant numbers of waders feeding and roosting on these estuaries, including Curlew, Redshank, Dunlin and Oystercatchers.

The island of Anglesey, barely 30 miles across, coastally is like Wales in miniature. Its shoreline, which includes the rugged Holy Island, has every kind

Sandwich Terns

of marine habitat; dramatic cliffs, sandy bays, salt marsh, dunes and small islands. Not surprisingly, Anglesey holds a correspondingly large variety of breeding sea birds and other species with an affinity for the coast. Here on this island county are to be found most of Wales' breeding terns. There are just three or four tern colonies, containing a variety of species in different combinations. At Cemlyn Bay, 400 pairs of Sandwich Terns breed on grassy islands in freshwater lakes close to the shoreline. With them are about 20 pairs of Common Terns and a few Arctic Terns. If you visit this wardened reserve, be very careful to avoid walking along the ridge of the shingle bank and putting the colony to flight. At the Rhosneigr colony on the opposite side of the island the population statistics are almost reversed. There are 250 pairs of Arctic and some Common Terns, but few Sandwich Terns on this RSPB reserve.

With the exception of the Black Terns (which nest on marshes but often occur over the sea during migration), the terns share a lot in common in their general appearance, having snowy breasts, grey mantles and velvet black caps, with long slender forked tails. They are among our most graceful seabirds. Watching them hover over water to plunge head first, like a flight of arrows, to catch sand eels and other small fish is a truly impressive sight. The differences in beak colour are a useful means of identifying the various species. The largest one, the Sandwich Tern, has a black bill with a yellow tip, and the loose feathers on its nape give the back of its head a rather 'scruffy' appearance compared with the smooth, flatter heads of both Common and Arctic Terns. The Common Tern has a red bill with a black tip whilst that of the Arctic Tern is plain red. This tern has a grey tint to the breast which also distinguishes it from the other species. There used to be a large colony of this bird on the rocky Skerry Isles off the north coast of Anglesey but the birds have long abandoned the site.

The rarest tern is the Roseate, an endangered species which is almost totally confined to Britain and Ireland where there are fewer than 1,000 pairs. Most of these are now to be found on the eastern seaboard of Ireland. Numbers of Roseate Terns in Anglesey have fallen alarmingly from 96 pairs at three sites in 1989 to 45 pairs the next year, ten pairs in 1996 and only three the following year. Numbers of birds at the colonies can change dramatically and we must hope that the fortunes of this delightful species will quickly improve. The Roseate Tern is distinguished by a breeding plumage suffused with soft pink, a black beak and tail streamers which are much longer than those of other terns. Look out for them among larger numbers of other species at mixed tern colonies.

The only other part of the country which has breeding colonies of terns is the north-east corner of Wales which holds numbers of the Common Tern and the rare Little Tern. The former species is found nearer the head of the estuary while there are upwards of 150 pairs of Little Tern at Gronant, the only breeding site now in the whole of Wales. There is no public access for viewing the colony but birds may be observed travelling to and from the site from the bird hide at Point of Ayr a little to the east. This site, like the others, is rigorously protected since terns are prone to human disturbance and predation by foxes, Crows and even Kestrels. The Little Tern became scarce because it has the additional misfortune of choosing to nest on shingle beaches where it falls foul of the recreational habits of people and their dogs. The Little Tern, which was formerly much more widespread round our coasts, lays its three well camou-flaged marbled eggs in a scrape in the shingle. This, by far the smallest and most active of our terns, is distinguished by its yellow bill with black tip and white forehead, a feature not shared by other adult British terns in summer-time.

Oystercatcher

There are two other species, both waders, which nest on pebbly beaches. The Ringed Plover may be seen in springtime running up and down shingle banks as it strenuously tries to attract attention away from its nest. If this happens, it is best to move on and let the incubating bird return as soon as possible to settle on its clutch of four eggs. Both eggs and sitting bird melt into their surroundings since the ring pattern on the head and upper breast blends well with stone and shingle. The Ringed Plover still manages to survive as a breeding species on most stretches of the Welsh coast though it is absent from most of Pembrokeshire, south Ceredigion and south-east Wales. Otherwise it is one of the commonest waders, usually seen in small groups at all times of the year on sand or shingle beaches.

The second wader, the Oystercatcher, is of course one of our most familiar shore birds. Its black and white plumage and stout orange beak make it as unmistakeable as the Magpie. As its name implies, it feeds mainly on shellfish prised open by its powerful beak, so it is frequently observed probing among the rocks to which mussels, barnacles and other molluscs tenaciously cling. The Oystercatcher is a striking bird made even more conspicuous by its noisy piping trills and calls. Secrecy has no part in this bird's make-up. When sitting on the nest, a mere scrape in shingle or in the sand dunes, the Oystercatcher is all too obvious. If someone approaches it will tip-toe away from the nest and return when the 'coast is clear'. In parts of Ceredigion and other places where there is no suitable beach for nesting, it will nest on rocky ledges well clear of the incoming tide.

Gulls are an important feature of our coastal avifauna. Most nest on seaboard cliffs but the Black-headed Gull usually breeds in colonies on upland lakes while the Lesser Black-backed Gull also frequently chooses a site on level ground, sometimes well away from the sea. Before the trees grew into forest at Newborough Warren on Anglesey, there used to be a large colony nesting among the spruce saplings. Many Lesser Black-backs are found on islands where they usually nest on grassy slopes. Cardigan Island has a large colony and there are others on islands along the Pembrokeshire coast, the Severn Estuary and North Wales. The familiar Herring Gull usually breeds on inaccessible cliffs but it is not always fussy, and some have taken to nesting on chimney stacks in Newport, Cardiff and other towns in Wales. Not long ago, I was surprised to see a Herring Gull sitting on its nest in the middle of the 18th fairway at Borth golf course! A marker had been placed round the nest by club officials to protect the site. Whether the purpose was to protect the birds or the golfers I am not sure, since any club member who ventured too close was vigorously attacked by the indignant birds!

The Herring Gull is a fine bird in its pristine soft grey and white plumage, and evokes memories of childhood holidays by the sea as it wheels and glides over beaches and cliffs. It commands attention by calling noisily from a hotel roof or chimney stack or begs greedily for bread or cold chips from passers by along the promenade. Its size and red-spotted yellow bill give it an air of aggression as it jostles smaller gulls and crows aside in its quest for easy pickings.

Many birders, however, are less than impressed by the gull family. Nearly all of these birds are white in colour, with grey wings and mantle and black primaries. Juvenile birds of all species are mottled heavily with brown. They are scavengers at rubbish dumps and pick the leftovers of man's fishing and farming activities and predate the eggs and young of other birds - the marine equivalent of crows. Gulls lack the elegance and appeal of the terns which have similar colouring. Until the last decade or two, perhaps the most off-putting feature of this family, however, was that they were all considered to be common and easy to observe, leaving nothing to those elements of surprise and diagnostic skill which are so important in the lives of birders! As the realisation slowly dawned that this perception is far from the truth, more interest has been shown in this family over the past generation. But first, the common species.

The Great Black-back Gull is our largest gull and is easily identified by its huge size and charcoal back. The Kittiwake has black legs and coal black wing-tips with no white 'mirrors' in them. Both of these species nest locally on Welsh cliffs, the Kittiwake sometimes in quite large colonies. If you are looking at a mixed flock of gulls, look at the legs to help identify them. We will suppose they are adult birds, since the mottled or patterned first or second years present extra problems which are really best addressed at the open pages of a comprehensive illustrated guide! The large birds with pink legs will probably be Herring Gulls. The Lesser Black- back is a similar size but has yellow legs and its mantle is much darker (purplish grey) than that of the Herring Gull. The smaller Black-headed Gull has red legs (and bill). In winter it has a dark spot behind the eye but no black head. Similar sized birds with greenish legs (and bill) are Common Gulls, which are found in Wales mainly in winter.

All of these species are common and most can be seen everywhere, roosting on beaches, scavenging at rubbish dumps, or seeking offal cast from fishing boats at sea or in the harbour. This dependence on humans has enabled them to boost their numbers significantly, but recently the population of Herring Gulls was badly affected by botulism which has caused a sharp if temporary decrease in the species.

Picking out the rarer species from much greater numbers of abundant ones is not always easy, and sometimes requires not only skill but a keen eye and patient observation. The Mediterranean Gull for instance, may occur as a lone bird among hundreds of Black-headed Gulls. It can be picked out at all seasons by its lack of black on the wing primaries and by the white ring around the red eye, and its heavier deep blood red bill. Last March I drove past a flock of 150 Black-headed Gulls gathered in a flooded field, not expecting anything out of the ordinary, when an unusually pink flush on the breast of one bird arrested my attention. I hurriedly stopped the car to take a closer look. This bird proved to be just a variant Black-headed Gull but among the flock were no less than three Mediterranean Gulls. The Little Gull, reminiscent of a diminutive Black-headed Gull, has a small beak and shorter rounder wings which appear darker underneath. It occurs in small groups and can turn up at almost any time of the year. All of these three species sport black heads in breeding plumage, but in winter there is just a residual black mark behind the eye. Note the white fore-wing of the Black-headed Gull, which is an excellent identifying feature when the bird is in flight.

The large Glaucous Gull and the smaller but otherwise not dissimilar Iceland Gull are whitish birds which, like Little and Mediterranean Gulls, have no black primary wing feathers (a feature always worth looking for). They are in a league of much scarcer species but like the North American Ring-billed Gull are observed on at least a dozen occasions every year. The latter species has been recorded regularly at Black Pill near Swansea, one of the best places in Britain to look for it. The even rarer Sabine's Gull, Ross's and Laughing Gulls have all been seen in Wales, and are sought keenly by some specialist observers.

Mediterranean Gull (top) and Black-headed Gulls

The 'real' seabirds, so far as many people are concerned, are the auks, Gannets and petrels which breed on steep cliffs or remote islands and spend much of their lives far out at sea. The auks come ashore to breed in the spring and leave their nesting cliffs by July, appearing onshore again only after the severest of storms or periodically during autumn or winter. At New Quay Head in Ceredigion for

45

instance, you can see both Guillemots and Razorbills breeding as close neighbours on the same cliffs. Typical auks, they are plump, broad-beamed birds which fly on wings better adapted as under-water paddles than for aerial flight. They skim low over the water, pumping their short wings before plunging clumsily out of the air into the surging waves below. Razorbills are darker above than the sooty looking Guillemots but both species are white underneath. The Razorbills nest on the more sheltered cliff ledges while the Guillemots crowd together on the more exposed ones. One egg is laid and this would roll off the cliff edge into the sea were it not for its elliptical shape. These two species breed on many suitable cliffs in Pembrokeshire, Gower, Caernarfonshire and Anglesey. The Black Guillemots on the other hand, breed at only one site in Wales, at Fedw Fawr on the north-east coast of Anglesey, the most southerly site in Britain of this northern species. This guillemot usually chooses a more sheltered spot to raise its single young than that taken by the commoner bird.

The Puffin ranks highly on almost everyone's list of favourite seabirds, chiefly on account of its exotic appearance. Its coloured, parrot-like bill and its comic waddle on short red legs gives it an endearing quality. Like the other auks it uses its paddle-like wings and webbed feet to propel it fast under water, expertly pursuing the fish on which it feeds. Unlike the other auks it lays its eggs in burrows on hill slopes, not on precarious cliff ledges. For this reason the Puffin must nest in remoter less accessible places to avoid predation by rats and other mammals, and consequently almost all of our Puffins breed on islands.

The Fulmar is one of our most familiar seabirds, found nesting on sheltered ledges or steep grassy slopes overhung by slabs of rock. Here in Wales, it has spread to many suitable cliffs since it first bred in 1947. This colonisation was part of an ongoing process of expansion which has continued remorselessly since the species first ventured from St Kilda, its only breeding quarters in The British Isles until 1844. Watch the Fulmar swerve or circle close to the cliffs on stiff straight wings, which immediately distinguish it from the Herring Gulls and Kittiwakes that often nest on the same cliffs. The wings are darker than those of the Herring Gull with lighter patches at the elbow. You need to be very close to see the extraordinary pale blue tubular nose which looks as though it might have shattered in collision with a rock face.

To observe the Gannets or Shearwaters we must either visit their nesting colonies on islands off the coast or keep watch out to sea, where both species may be seen wandering far from their breeding haunts. The impressive Gannet commands attention with its six foot wing-span and snowy white plumage (with black wing tips) which contrasts vividly against a background of grey waves and overcast skies. Watch a group of them circling over schools of fish, and then, with a sideways twist, wings angled and folded to afford smooth entry, they plunge almost vertically into the depths of the sea with the supreme skill of an Olympic diver. On consecutive August bank holidays scores of Gannets and more than a 1000 Manx Shearwaters appeared off the coast of Borth, diving one after another into the brine in pursuit of the concentrated shoals of fish. There is a huge Gannet colony on Grassholm, 10 miles off the Pembrokeshire coast. This is one of the four largest Gannet colonies in the British Isles. The other three are on St. Kilda and Bass Rock in Scotland, and Little Skellig in Ireland.

There are large numbers of Manx Shearwaters too, nesting in their island burrows at Ramsey and Skomer in Pembrokeshire and at Bardsey in Caernarfonshire. The Skomer colony is currently estimated to number 200,000 pairs, making it the largest breeding colony of this species anywhere in the world! The Manx Shearwaters are much harder to pick out than the Gannet since the brownish black upperparts are often obscure against the ocean back-ground. Their white under-parts are only glimpsed in flashes as

Gannets

they glide and turn effortlessly over the waves, their long straight wings al-most skimming the surface. There are rarer shearwaters to be seen regularly in the seas around Wales, including the Sooty and Mediterannean Shearwaters (similar but with lighter upperparts than the 'Manx') and the larger Great and Cory's Shearwaters. Sea watching at this level requires expertise, patience and powerful telescopes.

The Storm Petrel, which is little larger than a sparrow, is one of our most interesting sea birds, partly because it is so improbably tiny compared with such giants as Gannet and Cormorant. Its nocturnal habits also make it one of the most difficult to see. Like the Puffin it nests in burrows or crevices in rocky hillsides. In Wales, it is more or less confined to the Pembrokeshire islands of Skockholm, Skomer and Ramsey, although isolated pairs or very small groups elsewhere might easily go unnoticed. There is a regular movement of Storm Petrels at sea during the summer well away from their nesting islands. Both this species and the larger Leach's or Fork-tailed Petrel, which breeds in a few colonies off northern Scotland, may occasionally be seen on-shore following storms and gales. The Storm Petrel flies in a much straighter fashion than the more erratic and larger Leach's Petrel.

The Cormorant and Shag are both found nesting on Welsh cliffs. The larger Cormorant is commoner and can be distinguished from the Shag by its white face and, in breeding plumage, by its white thigh patch. Both birds are expert at catching fish, and the Cormorant in particular is a very familiar bird in our bays and harbours. Diving repeatedly, it persists until at last it emerges with a flapping fish which it invariably swallows in one gulp. Frequently after swimming, groups of them can be observed drying out on the rocks with their wings held outwards, as if suspended on coat hangers. Shags are smaller more slender birds, and the area round the mandibles is noticeably mustard yellow on a bird with totally greenish-black plumage. The Shags may nest singly or in small groups while the Cormorant nests in dense colonies whose presence is detected at a distance by the rocks white-washed with excreta from the nesting birds. The largest Cormorant colony is on St Margaret's Island close to Tenby, but the most unusual one is at Craig yr Aderin near Tywyn. Four miles inland from the coast, it is the only inland colony of this species in the whole of Wales. After July most of our seabird colonies are deserted when adults and young disappear into the vastness of the ocean, but Cormorants and Shags feed fairly close to shore and rarely stray far out to sea.

Skuas do not breed in Wales but they are not infrequently seen on migration in spring and again in late summer and autumn from July onwards. Timing and place are important when looking for these exciting and agile birds, since they are usually detected well out to sea where they follow the migration of terns. Skuas are pirates of the seas, chasing terns and other seabirds relentlessly and unmercifully until their victims disgorge their fish. The large and powerful Great Skua can even intimidate the Gannet in this way. Those coastal promontories which jut out into the ocean far enough to intercept migration routes of seabirds are among the best places to look for skuas. Point Lynas on Anglesey, Bardsey, and St David's and Strumble Head in Pembrokeshire are some of the best locations. Arctic and Great Skuas are by no means exceptional, but the Pomarine and in particular the Long-tailed Skua are rarer. The former species is regular at certain times of the year at Lavernock in Glamorgan, Strumble Head and one or two other locations.

Although this more or less completes our round up of seabirds, there are a few other species to be seen regularly in coastal waters in spring. Most of these are winter visitors such as divers, grebes and duck which linger into spring long after most of their kind have gone, waiting for their northern lakes to warm up. Some non-breeding birds, notably the Common Scoter and Eider, may stay even longer. These two species of

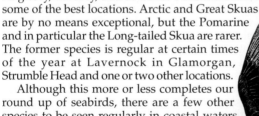

Storm Petrel

48

ducks can be observed around the Welsh coast at places like Aberdysynni in all months of the year, and in 1998 the Eider bred on Puffin Island, the first ever record for Wales. There are also other duck, apart from the abundant Mallard, which may be expected along the estuaries at all times and especially during the summer. The Shelduck is a familiar conspicuous bird with its tortoiseshell plumage and pink, knobbed beak. It usually lays its eggs on a lining of feathers in a tunnel, often an abandoned rabbit burrow. I was intrigued to see several some while ago, standing like sentries outside their nesting burrows beside a sandy embankment where open woodland fringed some fields more than half a mile from the coast. The Shelduck is usually to be seen feeding on the salt flats or swimming along muddy channels by the estuary, but it often makes its nest some distance from the sea. The Red-breasted Merganser habitually fishes in river estuaries and is often observed swimming close to the shore on the open sea in north and west Wales. Any duck observed diving in these habitats in summer stands a good chance of being this species. Its slim form low in the water, is quite characteristic.

Many of the birds most easily seen by the coast in summer are not seabirds at all. The estuaries, their salt flats and marshes attract Curlew, Redshank, Dunlin and on migration during spring, species like Whimbrel, Common Sandpiper, Black-tailed Godwit and Grey Plover on their way north. At this season, although the total number of wintering waders is small, a new summer coat may dramatically improve the appearance of those which linger into spring. The Grey Plover is a beautiful bird when its bland winter colours are replaced by its white-bordered black face and belly to match its dark-flecked silvery grey wings and back. In summer the underparts of the Black and Bar-tailed Godwits turn slowly from winter white to warm red-brown. In a flock of godwits you may see a continuum of plumage variation from winter colours through to full summer splendour at this time of the year. The Spotted Redshank, longer in beak and legs than the slightly smaller Redshank, and lacking the latter's conspicuous white wing bars in flight, is a rather average-looking bird in winter. In summer its dark dusky plumage transforms it into one of our most distinguished waders.

Towards the latter end of the summer from about July onwards, Greenshank and the smaller and much scarcer Green Sandpiper are among the typical non-breeding waders migrating south. At this time they are often found beside estuaries as well as rivers, lakes and reservoirs, but more often near the coast than inland. In flight both show conspicuous white rumps contrasting with dark back and wings. The rarer Wood Sandpiper is best told from the similar Green Sandpiper by its paler wings which are browner above and mottled and much lighter on the underwing. To a lesser extent, these three species may also be encountered on spring passage.

Last September I stood where the tidal river meets the sea at the Gann, Pembrokeshire comparing two small waders which were busily probing in the mud. They stayed so close to each other they could have been bosom friends. Both were neat birds with blackish legs and beaks, but one was much smaller than its companion. The larger bird with slightly drooping beak was a Dunlin. The other was the diminutive Little Stint, the smallest of our waders and little bigger than a sparrow. Eventually they flew off rapidly together upstream with barely a few centimetres of air space between them. Not far away, at the far side of a shallow pool, a small group of waders were also actively feeding on a sandy shore in front of some aquatic vegetation. At a distance they looked like Dunlin but on closer inspection they had a more elegant, taller and less squat appearance, with a hint of rose peach about the neck. The distinctly decurved bill immediately settled their identity as Curlew Sandpipers. Had they flown, their white rumps would have been a further distinction from the common-place Dunlin. Both Stint and Curlew Sandpiper(and probably the Dunlin too!), would no doubt have moved on within the next two or three days.

There are several species of terrestrial birds which are commoner on the coast than elsewhere. The purplish Rock Pipit is larger and darker than the Meadow Pipit which may be encountered on the same grass-covered cliff slopes. The Rock Pipit is totally confined to its marine habitat where it makes its nest in a rock crevice or steep bank and can be regularly observed searching for flies among the rock pools. The colourful Stonechat is more easily observed along the coastal footpaths than elsewhere, while in winter it is a familiar bird on coastal marshes and low lying uncultivated fields. Watch the male bird clicking loudly from a fence or atop a gorse bush. His chocolate brown head, white collar and peach-red breast make identification easy, particularly as the birds make no attempt to skulk in dense vegetation as so many smaller birds do. The Stonechat benefits from the mildness of the marine climate in Wales and the abundance of gorse in which the pair invariably make their nest. Some pairs frequent inland hill slopes and commons where there is a profusion of heather and gorse, but there the species is not nearly so common. Stonechats are partially resident birds which suffer badly in harsh winters, which explains why they are most numerous in the western, maritime parts of Britain. Some wiser individuals actually migrate to the Iberian Peninsula in winter.

Stonechat (male)

If you are by the coast in a habitat dominated by gorse, the small bird you are likely to see most often is the Linnet. It is found inland as well, on commons and scrubby farmland but there it is less abundant and declining in numbers. Listen to its characteristic twittering sounds until you are familiar with them. The Linnet has whitish wing bars, pale rump and warm brownish colours. Only in bright light does the crimson on the male bird's forehead and chest flicker clearly like a flame. In Wales the Common Whitethroat is another species which seems more at home close to the coast than inland. It may be found in close company with the Linnets if there are ditches and low overgrown hedges, brambles and tangles of vegetation. Unlike most of our warblers the Whitethroat will utter his rather scratchy song from an exposed post or the top of a hedge where his grey head, white throat and warm brown back and wing feathers will readily clinch his identity in good light.

The Raven and Peregrine both take advantage of cliff nesting sites along the rocky stretches of the Welsh coastline. Indeed, probably nearly as many Peregrines breed there as they do at inland sites. Pembrokeshire is particularly well endowed with them and there are many eyries on coastal cliffs in other counties like Caernarfon and Ceredigion. At Aberystwyth a pair can sometimes be seen flying over the town or chasing pigeons out at sea, and a pair of resident Choughs can often be spotted over the same cliffs as the falcons at Constitution Hill. A good walk along the coast in this part of Wales will almost certainly produce Peregrines and Choughs which, happily, seem to be getting commoner along the Welsh coast. In early May last year I noted a flock of 25 Choughs near Llangranog in south Ceredigion, and flocks of this size are becoming a more familiar sight.

Choughs like steep cliffs with crevices or caves in which to build their substantial Jackdaw-like nests. They alone among the crow family feed exclusively on insects and worms which they obtain on the bare and bracken-clad slopes and fields which typically adjoin their nesting cliffs. Once you are used to them, Choughs can be picked out from other 'black crows' without too much difficulty. At a distance they look very black compared with the duller Jackdaws which also nest in colonies (often in burrows) along suitable cliffs. When they fly overhead or below the cliff the splayed out fingers on the wing tips are seen and the wings are broad, almost moth like. Look for the needle-like curved red beak. This is the ultimate feature to clinch any identification. With experience the shape of the bill is enough to identify the Chough even when the colour cannot be seen. The long drawn out 'chough' sound (which unlike the name of the bird rhymes with 'bough' not 'buff'), echoes along the cliffs and immediately attracts attention. Often the Chough will let you approach it quite closely as it probes the ground for grubs, since it is not a shy bird. In flight it swoops upwards, or dives on folded wings which is quite characteristic and unlike other members of the crow family. Jackdaws are smaller, their wings are smoother, rounded and beat more quickly. At close range the grey nape is a diagnostic feature of the Jackdaw.

The greatest numbers of Choughs in Wales are in Pembrokeshire and Caernarfonshire with smaller numbers elsewhere. The population is increasing and this special Welsh bird, after resuming breeding on Anglesey in 1960, has now returned after a long absence to the cliffs of Gower, its most southerly breeding station in Wales. The Chough, in fact, is only found in Wales on the British mainland, although there are some on Islay in Scotland, the Isle of Man, and many more in Ireland which holds the bulk of the Chough population in these islands. The total Welsh breeding population currently stands at about 170 pairs. The roost at Craig yr Aderyn in south Meirionnydd now reaches a peak of almost 80 in some autumn months.

There are other land birds which benefit from cliff or stony sites for nesting. Wheatears often nest under boulders above the tide-line and Kestrels and Rock Doves choose sheltered cliff ledges to lay their eggs, although neither makes any attempt to build a nest beyond placing a few sticks. The Domestic Pigeon was originally bred from Rock Doves, and tame and wild birds are so interbred that there are few genuine Rock Doves left in most parts of Britain. The classic features of truly wild birds include two thin black wing bars and a white rump. Flocks of feral pigeons are easily told from wild pigeons of all species since they usually display wide variations in plumage between one bird and another, and some individuals at least show a considerable amount of white in flight.

Most of the 'land' and 'waterside' birds described in this chapter can be found without too much difficulty, but the coast also attracts some of our rarest migrants and accidental visitors during spring and summer months. Some rare warblers are recorded regularly, species like the Icterine and the Melodious Warbler, two similar green and yellowish warblers of the 'Hippolais' genus. The Subalpine and Barred Warbler are other summertime 'regulars', while early autumn may bring the Yellow-browed and Pallas's warblers. Related to the Willow Warbler the former species has two diagnostic wing bars while the tiny Pallas's Warbler, only the size of a Goldcrest, sports not only the double wing bar but also a distinctive striped crown. Most of these species are recorded on headlands and islands, outstanding of these being Bardsey and Skokholm. Headlands in Pembrokeshire, the island of Skomer, Kenfig Pool in Glamorgan and South Stack on Anglesey, are examples of other places which offer a reasonable chance of a rarity: a Woodchat Shrike in June, a Rose Coloured Starling in September or a Scarlet Rosefinch at almost any time of the year are all possible. These are all well-watched nature reserves where skilled ornithologists using good telescopes and mist nets claim many of the unusual records. It seems fair to assume that, given fuller coverage elsewhere, many of these rarities would be recorded more often in other parts of Wales.

THE COAST in SUMMER - species to look for:

GENERAL	OUT AT SEA	ESTUARIES
Herring Gull	Terns	Shelduck
Black-headed Gull	Skuas	Red-breasted Merganser
Great Black-backed Gull	Gannet	Mallard
Lesser Black-backed Gull	Petrels	Curlew
Oystercatcher	Shearwaters	Whimbrel
Cormorant	Eider	Redshank
Heron	Common Scoter	Lapwing
		Ringed Plover

CLIFFS (sea birds)	CLIFFS (land birds)	BEACHES
Fulmar	Peregrine	Gulls & Terns
Shag	Kestrel	Oystercatcher
Guillemot	Buzzard	Curlew
Razorbill	Jackdaw	Whimbrel
Puffin	Chough	Sanderling
Kittiwake	Rock Dove	Dunlin
	Wheatear	Ringed Plover
	Whitethroat	Pied Wagtail
	Rock Pipit	Rock Pipit
	Meadow Pipit	
	Linnet	

Cormorant

Notes: (1) The species listed under 'general', together with Shelduck, Mallard, Curlew and Dunlin which are all mainly associated with estuaries, are seen everywhere along the coast in suitable habitat.
(2) Gannets, Storm Petrel, and Manx Shearwater nest mainly or exclusively on islands, but occur widely elsewhere especially after stormy weather.
(3) These lists are incomplete and exclude many species which might occur by our coasts in spring or summer, such as migrants and lingering winter visitors.
(4) There is a strong overlap between species normally found on beach and estuary since stone, shingle, sand and even mud may be found on either.

Guillemots (top)
and Razorbill

CHAPTER THREE
MOOR AND MOUNTAINS

The highest peaks of Wales and the most rugged landscapes lie in the north of the country, mostly within the Snowdonia National Park. Here we can truly speak of mountains, whose high tops on wet days are invariably hidden by a mysterious veil of clouds. When the mists lift above the mountains a spectacle of pointed summits, forbidding grey cliffs, and awesome boulder strewn passes are revealed. Deep black lakes nestle between the stern hills, whilst near the summits still, dark pools lie eerily beneath the topmost crags. On clear days, or more likely, during clear spells in this rain drenched region, the panorama is changed completely in tone and colour. The mountain slopes soar upwards in breathtaking fashion to meet high corries and hazy blue peaks. The glass-like surfaces of the larger lakes mirror the reflected green, stone-grey and purple of grassy foothills, heather-carpeted inclines and towering cliffs. For the onlooker, gazing at such inspirational scenery and drawing deep draughts of mountain air, this is a time for contemplation and spiritual uplift, leaving worldly cares below. No wonder people don their walking boots and weather-proof clothing, drawn by the magnetic attraction of these mountains.

Highest among these alpine hills is Snowdon, a well shaped mountain which at 3560 feet (1085meters) is the highest in England and Wales. Since 1885 refreshment at its summit cafe has been made available to all during the tourist season by the steam mountain railway which takes passengers effortlessly from the station at Llanberis, all the way to the top. To the east of Snowdon, the rugged Glyder and the Carneddau plateau have several peaks above 3000 feet which are accessible only to those committed to strenuous exercise.

No other mountains in Wales reach 3000 feet although the Arans and the legendary Cader Idris in Meirionnydd come close to it, while further south, Pen y Fan in the Brecon Beacons reaches more than 2900 feet (886 metres). Other notable mountain ranges include the Rhinogs, the central Cambrian chain of hills, the Berwyns of north-east Wales and the Black Mountains in the south. Even in these ranges the terrain is, for the most part, often better described as hilly rather than mountainous. There are many impressive crags and deep gorges, but the summits of those hills, at, say, between 1500 and 2500 feet are more often than not gently rounded and usually covered with heather or coarse grasses which cling to thin peaty soils and shale. Boulders and scree may surround the highest points, while a pile of carefully placed stones often denotes the collective achievements of generations of walkers in reaching the summits. Gazing around from this triumphal point the hiker may look upon a peaceful scene, often with no other soul in sight, not a house nor other sign of human habitation. Undulating green hills bordered by old stone walls drop into deep vales, and wild remote moors stretch into the distance. Sometimes a winding track leads past a farmstead surrounded by wind-blown rowan or hawthorn trees, and this route onto the moors may well be the starting point for a day's birdwatching in the hills.

The moors will be revisited later, but first let us start in the spectacular setting of the highest mountains, although we must begin on a faint note of dis-

appointment. There are no species in Wales which are confined to these peaks and plateaux, no breeding Snow Buntings nor Ptarmigan such as we find on the highest hills of northern Scotland. The Cambrian mountains, it seems, are just too far south to attract these species which are in any case at the southern end of their range in the Scottish Highlands. The majestic Golden Eagle and the even larger White-tailed Eagle are but rare wanderers to Wales, although both species probably survived here two hundred and fifty years ago. But the highest ground is well endowed with crags, scree slopes and deep ravines and in this habitat several species benefit from nesting ledges inaccessible to all but accomplished rock climbers, and crevices well protected from the frost, rain and biting winds.

Raven

The Raven lays its eggs at the end of February or early in March while the weather can still be severe on the high hills. The nest in this setting is often a huge structure placed on a ledge beneath a protective overhang, or wedged into the crevice of a cliff face. Overhead the courting pair honk like grunting pigs, and circle in soaring flight displaying their distinctive wedge shape tails, or roll over in their astonishingly unique way and fly briefly upside down. The Raven seems to embody the spirit of the mountains, as its guttural calls echo between the hills. Yet the majority in fact nest in trees in pine and deciduous woodlands, whilst others inhabit shoreline cliffs. On the most barren moors, Ravens will sometimes make their nest in a low bush in the absence of a more suitable site, just as Carrion Crows regularly do in stunted trees in upland gullies and near coastal cliffs.

The Chough emulates the larger, carnivorous, carrion-eating corvids by also nesting in the Wagnerian setting of the mountains of North Wales. Yet the insectivorous Chough is really less suited to such hardships and thrives best under more clement conditions. Abroad it is a familiar sight in some parts of Spain and other countries of the Mediterranean. The Choughs of Caernarfonshire are the only mountain dwellers of their kind in Britain except for a few which until recently inhabited the crags of Mid Wales. You can see them on the slopes of Snowdon, or near the pyg track at the top of Llanberis pass, where their sneezing calls resound among the canyons.

The Peregrine is able to find more choice nest sites among the gorges and steep cliffs of Caernarfonshire and Meirionnydd than in any other inland areas of Wales. In districts further east where suitable nest sites are at a premium, pairs habitually nest in quarries and often seem undeterred by working activity provided the human presence is not too intrusive. Peregrines frequently nest alongside Ravens, occasionally occupying the latter's disused nest in April. When disturbed, the falcon will emerge from the face of its nesting cliff, chattering boldly in a mixture of anger and alarm, although I have seen a female quickly return to her eyrie not far from where climbers were exercising nerve testing skills in scaling a sheer rock face. They were too intent on their footholds to even notice the Peregrines!

Ornithologists are deeply concerned about the fall in numbers of the Ring Ouzel or Mountain Blackbird as it is otherwise known. I can remember them 10 or 15 years ago occupying several sites which are now deserted. In the mountains of the Snowdonia National Park is the best area to look for them, though there are others still clinging on in the Elan Valley, the Brecon Beacons and other rocky places in Wales. These days you may need to climb to the higher crags to find one. Ideally they like steep, heather-covered gullies in rocky ravines with a scattering of trees. The Blackbird-like nest is usually on the ground in heather, in a bush, or in vegetation covering an outcrop of rock. Listen for a musical, fluting and thrush-like song of three or four notes echoing from the hillside in May or June. The birds may be hard to pick out at first among such a vast terrain of rock and heather, and you may have to wait until they fly or flit between rocks before you spot them. The 'chack-chack-chack' call resonates loudly but you may spot the male singing from the pinnacle of a rock or branch of a small tree. His blackish feathers are flecked with white but the crescent shaped white throat patch is the clearest feature. His mate is rather browner and her throat patch is less distinct but otherwise the two sexes are quite similar.

Where have all the Ring Ouzels gone during the past two decades or so? The cause is not clear but when numbers are low, residual populations may be more vulnerable to predation. Since the

Dotterel

Peregrine has increased, the Ring Ouzel must now often share its rocky domain with a pair of falcons perhaps nesting only 200 or 300 metres away. Observers have noticed a correlation between the arrival of Peregrines at new eyries and the disappearance of this species. The Ring Ouzel, the same size as a Blackbird, would make a reasonable meal for hungry young falcons. The male sometimes perches quite prominently when singing during the breeding season and it requires only one or two sudden and fatal strikes during the two months in which the two species share the same cliffs to eliminate the Ouzel from its haunts. There is some evidence, though, that some pairs have learned to counter the dangers posed by falcons by nesting in more sheltered sites such as at the perimeter of conifer plantations.

The Dotterel is one of the most beautiful and attractive of British birds. Once or twice it has been found nesting on the highest grassy upland plateaux of Snowdonia, but it can easily be missed in remote mountainous country. In Scotland its breeding numbers were greatly under-estimated because of this. There are perhaps 700 or 800 pairs on the munros (over 3000 feet) in the Highlands instead of the 70 or 80 pairs originally estimated. If you are walking on the high hills, keep your eyes open for a medium sized plover with a combination of grey upperparts and a bright ruddy chestnut breast and dark underbelly. It has a decorative dark-bordered white band above the breast and a white eyestripe. A bird beautiful enough to enliven any bleak stony ridge walk, the Dotterel can be quite tame. One perched on its nest within feet of me one June when I was walking in snowy weather on the Cairngorms. Your best chance of seeing a Dotterel in Wales is during migration at the end of April, early in May or in late summer. Try the summit of Plynlimon or Tair Carn (Carmarthenshire) where they are seen in most years, or if you are fit enough for a longer hike you may see them on the Arans in South Meirionnydd. Plynlimon ridge, the highest place in Ceredigion, is an easier option which is quite popular with walkers who can reach the summit in about one and a half hours following a gentle slope from Eisteddfa Gurig. An alternative route along heather tracks near Dylife takes you through the Montgomery Wildlife Trust reserve at Glaslyn. There are still a few pairs of Golden Plover and Red Grouse in the area and you may possibly come across a Ring Ouzel.

Further to the south, the Elenydd district of the Cambrian mountains provide refuge for most of the Golden Plover and Dunlin now breeding in Wales. A 1996 survey of 150 square kilometres of the Elenydd conducted jointly by the RSPB and the Countryside Council for Wales, discovered 43 pairs of each species. These figures though are likely to be rather lower than the true numbers, especially in the case of the Dunlin, which can be very difficult to locate. Often enough we have visited their territories in spring and been unable to detect any sign of these birds which we felt sure were concealed somewhere on the peat bogs of these uplands. In order to see or hear these birds, a very early start is likely to be more effective than a later one, unless, as the naturalist William Condry suggested, a sleeping bag is taken and a night is spent on the hills. At such inconvenient times moorland waders such as Dunlin and Curlew are more vocal. The male Dunlin utters an accelerating trill which rises and then falls, lasting about four seconds. These are the notes to listen for when surveying a likely area for Dunlin in May.

At first sight, the views from the round-topped hills look unpromising, even on clear days; a desert of arid green and ochre moors consisting of sheep's fescue, matgrass, bents, juncus and other grasses, broken occasionally by a rocky outcrop. Trudging through this lumpy saturated ground can be heavy going. Often you have to retreat, retracing your steps to find a dryer safer route since the peaty ground can be treacherous. On wet or cold windy days, one is left wondering whether it would have been wiser to stay at home. There are usually a few Crows, Ravens and Buzzards on these hills searching for carrion. The only other species may be the stoic little Meadow Pipit, emitting its plaintive thin 'seep, seep' note as it tosses in the breeze, or the Skylark, showering the desolate hills with its torrent of mellow notes from high in the sky. Here at least the Skylark can escape the ravages of chemical fertilisers, insecticides, farm machinery and the tramp of hoofs.

An hour's walking may produce little change in the monotonous scene. Then, wearily reaching the brow of the hill at the top of a long slope, suddenly the bog lies ahead of you. Eroded hummocks of black peat stand upright like sentries, capped with heather and encircled by treacherous pools of deep brown water. Among acres of bog the heather is secure, safe from the grinding jaws of sheep which avoid such hazardous ground. Here you walk with care, listening attentively for the sound of something different. The imitative trills of a Skylark above can cause a moment of false expectation. It is a wonderful mimic and can copy notes of Dunlin, Curlew and other species, and is more likely to be miss-identified as a rarer bird when the ears are strained in anticipation of something more exciting. Last season I surveyed an ideal upland bog but heard nothing except the rattle of a Red Grouse courting its mate among the sparse patches of heather. Then, when I had given up and began walking back up the sodden slope of a hill carpetted with stunted heather, my squelching boots put up a single Dunlin. In such a desolate expanse this bird looked like a sparkling jewel. I was close enough to observe its black eye and matching dark, slightly decurved bill which gives it a rather comic appearance. The bird ran a little way and then walked, enticing me further from its nest which I presumed was nearby. I was more than happy to comply since I knew she would return to her clutch of four eggs as soon as I was a safe distance away.

The Golden Plover is rather more obliging than the Dunlin in revealing its presence since its melancholic piping note is usually given from a slightly raised hummock of grass. It is rather like a radio-bleep, aiding the observer to accurately locate the calling bird. Listen for the Golden Plover and when you hear its plaintive piping call, scan the hummocks until the bird is focused in your binoculars. Often the plover will be spotted standing on a tussock of moss-covered vegetation and will draw you away from its nesting area. The golden crown, wings and back flecked with dark edgings contrasting with the white bordered black face, chest and belly, make this plover one of the most lovely waders in summer plumage. Once you've enjoyed a view of the male bird guarding his territory, it is time to move away since the female will leave her nest if you approach any closer.

The undisturbed islands of heather between the black acidic pools, the reedy fringes and more varied vegetation of the upland bog, provide habitat for a number of other interesting species. A Snipe will often be heard drumming

Dunlin at nest

overhead as it dives, splaying out its tail feathers to produce this distinctive mechanical sound, or more often its 'pumping action' double notes will be heard from the concealment of the reeds. The Red Grouse is not a common bird on these moors but you can usually come across one or two pairs where there is enough heather. If there are pools fringed with reeds, Teal may still be seen foraging for food near the water's edge. Ceredigion used to be a good county for Teal but unfortunately they too seem to be declining. The Elenydd Survey found only eight pairs, and of these just one pair was proved to breed.

Anywhere there are boulders, scree slopes, or even stone walls, at elevations high or low, the Wheatear is sure to be found. It is a greyish handsome bird which shows a conspicuous white rump as it flits between rocks and boulders. The female and juveniles are a little more sandy in colour than the male who is blue-grey above with a distinctive black patch around the eye. Both sexes have white underparts and black wing primaries. The Wheatear finds plenty of crevices or holes in the ground under rocks to protect its nest and five pale blue eggs. Most semi musical notes heard from distant rocks and corries, when tracked to their source, will prove to be delivered by a Wheatear. Like the other chats the Wheatear is often first detected by its clicking notes, which sound like two stones being rubbed together. More musical thrush like notes, unless sung by a Blackbird or Mistle Thrush, may lead you to a Ring Ouzel.

The best and most extensive true moorland, blanketed with ling and deep heather, is now to be found in north-east Wales. The Berwyn mountains provide ideal habitat for the 25 pairs of Hen Harriers which grace these uplands. The male bird is a wonderful sight; his soft grey and white plumage with black wing tips makes no concession to concealment as, like a ghost, he quarters the moors for birds and rodents. With luck and a little patience, you may see one from the road which passes through the RSPB reserve on the border between Montgomeryshire and Clwyd. The Hen Harrier lays its eggs deep among the surrounding heather where unfortunately it is predated heavily by both foxes and some game keepers who consider it a serious threat to their grouse stocks. The Red Grouse is in fact increasing on these moors and there are now about 100 pairs in and around the reserve. Elsewhere in Wales the Grouse is patchy in its distribution with just a few pairs here and there in most upland areas. Numbers are largely determined by the abundance of heather which unfortunately is losing out to grazing land. The Radnor Forest, Black Mountains in Breconshire and the Brecon Beacons hold reasonable if declining numbers of this species. Familiar to most people, the Red Grouse is the epitome of moorland birds and gives a particular pleasure to those who come across it during a long trek in the hills. A plump, rusty-brown bird with a striking red wattle above the eye, its guttural rattling voice breaks the silence and breathes life into the hills on a misty or overcast day miles from human habitation.

The other grouse, the Black Grouse, is less familiar in the public mind but is perhaps a more fascinating bird. We will come across it again in the next chapter, since in Wales it is usually encountered in young conifer plantations, whose spread after the war resulted in a welcome increase in this species. Its most productive habitat is a combination of heather and bilberry moorland, young spruce plantations with birch and patches of grassland. I was a little surprised then some while ago to hear a lekking male calling from moorland almost devoid of trees, although admittedly the area contained the other requirements of mixed heather-moor and farmland. If you hope to hear this species an early

Curlew on nest

start is essential. We had arrived on site well before seven in the morning, and could detect the bubbling notes while still a quarter of a mile from the lek.

The Merlin, like the Hen Harrier and the Black Grouse, also inhabits the heather moors of North Wales where it usually lays its clutch of eggs in deep heather. Formerly, when the species was more prevalent in Wales, it also bred in sand dunes and other coastal habitats. In districts further south where its haunts may almost be devoid of heather, it has adapted to breeding in conifer trees. This fact is perhaps less surprising when one remembers that it habitually does so in North America and in the vast Taiga (conifer forest) which traverses Scandinavia and Russia. Heather nesting Merlins in Britain therefore, are the exception rather than the rule throughout their entire range. Both Black Grouse and Merlin, and to some extent other species like the Goshawk, make use of a mixed habitat of moorland and conifer forest. It seems to follow naturally, then, that the next chapter of this book should be devoted to the conifer forest but before we leave the open moors there are a number of other species to consider. Most of them are quite common but the inconspicuous Twite is in a different category, since there are just a few pairs breeding on the heather slopes of North Wales. Rather like a plain Linnet with corn-coloured beak and pinkish rump, this species is more familiar to most birders as a winter visitor when it is not uncommon in small flocks, usually near the coast. In parts of Scotland and the Pennines Twites are more plentiful, and the species may well be overlooked in Wales.

As we begin to leave the hills and slowly descend the slopes into rough pasture, there are interesting species to be found among the hummocks of coarse grass and bracken. This marginal land attracts Meadow Pipits and Skylark of course, but the Whinchat will be of more interest. This lovely bird may be observed singing from a fence-post, small bush or frond on bracken-covered slopes or marshy fields. Like the Stonechat, little attempt is made by the male on territory to conceal his position as he sings from an exposed perch but he may still be difficult to pinpoint in vast tracts of rough grass. Like the other chats, the birds are often detected by the clicking notes from which their names derive. After feeding, the incubating female will cling patiently to a swaying reed or stem of grass and wait till the coast is clear before she drops onto her nest containing five or six blue eggs, well concealed in the long grass.

Where the ground is damp and soft enough for the long probing bills of Curlew and Snipe you will find both of these waders, or would have done until the past decade. The Curlew is our largest wader and has the most extraordinary long curved bill. On the moors it immediately attracts attention as it calls excitedly over its nesting territory or trills delightfully as it glides down to alight gently on the ground. Like the Grouse, the Curlew is an essential part of our moorland heritage, a personality which brings sparkle and gladness to the hills. Its repertoire of calls rank among the most beautiful and evocative of all birds. The Lapwing, with its long crest, black and white plumage and wheeling flight on broad, rounded wings, is equally one of our most attractive birds. Its natural habitat is the wet meadows and soggy short-cropped marginal land, leaving the deeper coarse grasses to the Curlews. Its 'pee-wit' calls as it dives over its nest - a scrape containing four mottled eggs, are familiar to most people who are old enough to recall spring days in the countryside 30 or 40 years ago. Unfortunately, both Curlew and Lapwing have declined at an alarming rate. In northern England both species can still be found in good numbers in marginal land in the Pennines, but in Wales they are becoming a rare sight in such habitat. The Lapwing in particular has almost disappeared. The Snipe is also losing ground and may have particularly suffered from a series of dry summers, drainage schemes, and a huge extension of sheep grazing. Some people would argue though that the main problem for ground nesting birds is the increase in Crows and foxes, which in the latter case take not only eggs and young, but the incubating birds as well. This may well be true, especially since a reduced and degraded habitat will make the sitting birds much easier to find.

In those valleys where the hills descend steeply into bracken slopes interspersed with hawthorns, rowans and other trees, you can find more typical species. This kind of habitat, so common in Wales, is called 'Ffridd' in Welsh. Here there are many Willow Warblers in summer, whose descending melliflu-ous warbling is one of the commonest bird sounds heard in the countryside. The Tree Pipit has a very different, piping song. The displaying bird lifts off from a tree and rises in the sky before spiralling towards the earth, half folded wings held out in a parachute descent. The song continues throughout this display, tailing off as the pipit alights either onto another tree or onto the ground (the Meadow Pipit also has a parachute display, but this nearly always begins and ends on the ground). This kind of countryside is one of the best in which to find the declining Yellowhammer, especially where there is deep bracken and gorse with open clearings. With the sun reflecting the bright yellow of his plumage the male looks a lovely bird as he sings from an exposed perch. Often the song is heard before the bird is seen - a series of wheezing short notes culminating in a long drawn out one. The Redstart is usually abundant in this kind of habitat provided there is a good scattering of trees.

The Meadow Pipit seems an unlikely host for the robust and greedy young Cuckoo, but nevertheless it is one of that species' favourite victims. The Cuckoo, that harbinger of spring, is not especially common but the fringes of the moors seem to be among its favourite haunts in Wales. Here its familiar calls resonate across the hillsides and valleys. Although more often heard than seen, the Cuckoo is a largish bird which looks rather like a Sparrowhawk. The adult is bluish-grey in colour with pointed wings and long tail and usually flies lower and straighter than the hawk.

Given the extent and richness of the Ffridd on the perimeter of the Welsh hills, it is not surprising that such species as the Whinchat, Tree Pipit and Redstart are represented in such good numbers. But don't bother to traverse the hills for these birds in winter. By then they are all settled in their quarters in the Mediterranean or Africa. Even the Skylarks, Meadow Pipits and other small passerines have left the high ground and the Merlins which prey on them have followed suit. Conditions on the uplands in winter it seems, are just too harsh and food is too scarce for all except hardy grouse and scavenging Crows and Buzzards.

MOORS AND MOUNTAINS - species to look for:

GENERAL	DAMP-MARGINAL AREAS	SCARCE SPECIES
Meadow Pipit	Snipe	Goshawk
Skylark	Curlew	Peregrine
Carrion Crow	Lapwing	Kite
Raven	Skylark	Ring Ouzel
Black-headed Gull	Whinchat	Red Grouse
Pied Wagtail		Golden Plover
Buzzard	**FFRIDD**	Dunlin
Wheatear		Black Grouse
Curlew	Redstart	Merlin
Kestrel	Whinchat	Hen Harrier
Peregrine Falcon	Tree Pipit	Dotterel
Kite	Willow Warbler	Twite
	Yellowhammer	Short-eared Owl

Notes: (1) Most of the species listed 'general' may be seen readily in any of the upland habitats. The Kite is becoming a familiar occurrence in many parts of Mid Wales though still scarce elsewhere.

(2) Birds are few on the moors and mountains in winter. Several winter abroad (see check list for each species), others form flocks or move to the coast. Of the scarce species, the Red and Black Grouse and Goshawk are sedentary, most of the raptors and waders move off the hills, the Ring Ouzel migrates. The Dotterel is generally only observed on migration.

Wheatear

CHAPTER FOUR
THE CONIFEROUS FOREST

The Welsh coast, the wooded valleys and the mountains are places of great beauty, sought after by tourists and lovers of dramatic scenery. They are the subject of picture-postcards, calendars and travel books. No such claims can be made for the coniferous forests. At best their aesthetic benefits are cosmetic, softening the austerity of some of our moors and upland lakes which might otherwise seem barren uninviting places. Nevertheless, it is a pleasure to walk under fine stands of tall trees well spaced to allow bilberry and ling to carpet the forest floor, or green mosses and algae to creep along old stumps and fallen boughs. In winter a carpet of powdered snow can produce a pleasing contrast with the dark green foliage, especially when the snow clings thickly to the upper-sides of horizontal branches.

All too often the scene is very different; thickly-growing columns of stunted conifers devoid of light and leaf except for the topmost branches until the day they are either thinned or felled. The understory is a world of almost total darkness, and such plantations hold few birds except Goldcrest, Coal Tit and the usual Chaffinches and Woodpigeons. Trudging through these woods in late winter, you long to reach an open glade where warming beams of sunlight penetrate the chill air and brighten the forest floor. Your dedication will have been well rewarded though if you see Crossbills, Siskins, or some of the other species of the coniferous woods. Overall, I must admit to feeling a sense of excitement when bird-watching among the conifers, providing the sites are well chosen. Potentially there are interesting birds to be seen which we will be discussing in this chapter.

The coniferous forests are playing an increasingly important part in the conservation of wildlife in Wales. Already they hold the main concentrations of several species and important populations of a number of others. Furthermore,

Black Grouse at lek

Crossbill pair

Jay

Short-eared Owl

Hen Harrier pair

Kingfisher

Red-breasted Mergansers

Goshawk

Siskin (left) and Redpoll (both males)

the Forest Enterprise is working closely with conservation organisations like the RSPB to maximise the suitability of its forests for wildlife, while at the same time keeping in focus its commercial aim of producing timber. 'The Enterprise' also provides excellent facilities for the public including plenty of well signposted tracks, some of which offer splendid views between gaps in the trees to the mountain landscapes beyond.

Where acres of trees are clear-felled, the bare ground becomes attractive for Nightjars which need open heath and scrubland to hunt for moths and other insects of the night. In East Anglia, clear-felling has enabled the rare Woodlark to increase dramatically so that now the Brecklands are the principal stronghold of this species. Woodlarks bred in Wales until the mid-sixties but strangely they occupied a rather different, non-woodland type of habitat, so they are unlikely to take advantage of this activity. A six-fold increase in England, however, offers an encouraging possibility that this species may return to breed in Wales once again. A bird which has taken advantage of clear felling once vegetation has taken root is the Tree Pipit, one of the most characteristic species of the open areas of both conifer and deciduous woodland. Wider spacing of trees as they mature allows native vegetation like heather to return, especially along forest rides and woodland edge. The penetration of light to the ground encourages an undergrowth of numerous other plants to flourish. Stands of beech, birch, oak and other deciduous trees, usually planted on the fringes of the woodlands, improve the diversity, as does the planting of a wider variety of conifers. All of these measures taken together greatly improve the richness and variety of bird and other wildlife in the forests.

Some conifer trees are more wildlife friendly than others. The native deciduous larch is particularly good. Its well-spaced branches enable large raptors such as the Buzzard and Goshawk to land on their bulky nest platforms without risk of injury. The Crossbill may often be seen feeding on the seeds of larches as well as spruce trees, and Siskins and Redpolls seem to like them too. Provided the ground cover is sparse, with grass and mosses predominating, Wood Warblers are often plentiful beneath the canopy of these trees.

The development of forestry has directly led to a number of species becoming established as regular breeding birds in Wales. The Goshawk is discussed in more detail in a later chapter, but it can perhaps, best be observed displaying over its nesting woods in winter or early spring. Recently we watched a male displaying as early as December from a forest ride in a large mature plantation. The Goshawk circled and dived over its wood for several minutes, totally ignoring our presence on the path below. Sometimes he flew towards us, looking sharp-winged at this head-on position, rather like a Peregrine. Three times at least he gathered speed in a shallow dive and zoomed upwards in an exciting vertical climb. Once, as he turned away, we could clearly see the splayed white feathers either side of his tail, an integral part of his courtship display. Finally the Goshawk dived headlong into a sheltered part of the wood and half a minute later flew off out of sight. Out of curiosity we decided to see if the nest was close to the spot where he had dived into the wood. Sure enough, a large flat miss-shapen nest was there, close to a ride. We imagined the female, who did not put in an appearance, would refurbish it with a few fresh pine sprigs during the next three months, before laying in April.

The Crossbill may be seen at all seasons of the year in Wales. Nesting starts early in the new year when the forest floor may be carpeted with snow. To find Crossbills visit one of the larger upland forests, say, in January or February. Any flock, large or small, of sparrow-sized passerines flying in undulating fashion above the trees could very likely be Crossbills (smaller sized birds may be Siskins). The Crossbill is a rather heavy finch and when it lands near the top of a pine it has a dumpier appearance than the more slender Chaffinch. You may be able to discern the crossed mandibles as the bird hangs below a branch wrestling with a pine cone from which it expertly extracts the seeds. If you see him well, the red plumaged male is a striking bird, while the female and juvenile are greenish and striated with some yellow. Watching these birds the observer is forcibly reminded of diminutive parrots, although of course, they are totally unrelated.

The flight calls of the Crossbill are different from those of the Chaffinch and will help you separate them. The latter calls 'chip' or 'chip-chip', rarely more than two notes together, while the Crossbill pours forth a medley of higher octave 'chip-chip-chip' calls as it flies overhead or perches near the crown of a tree. Some notes of the Greenfinch are not dissimilar, but that species is rarely encountered in extensive conifer woodlands. The Crossbill has a variety of several other notes, and a flock of them may quickly attract attention as they feed in the topmost branches of a stand of firs.

Crossbills breed regularly if rather erratically in Wales following invasions from their Scandinavian and other European strongholds. These irruptions usually follow a poor cone crop in those countries from where they originate, forcing thousands of hungry birds to seek food supplies outside their normal range. Declining remnant populations may breed in Welsh forests for some time but eventually become extinguished unless there is a further influx. In Mid Wales for instance they have bred, though usually intermittently, in Radnor Forest, Tywi, Irfon and Crychan forests in north Breconshire, Strata Florida, Pen Dam, Cwmystwyth and the Artists' Valley in Ceredigion and the Dyfnant forest and Lake Vyrnwy in Montgomery. Further north, Coed y Brenin is one of their most regular breeding haunts and in fact they have bred in many areas where there are extensive plantations of conifer. They are usually absent even in 'good years' from parts of South Wales and from Pembrokeshire. In seasons when the spruce cone crop is sparse, the number of breeding Crossbills is greatly reduced.

The Siskin is another of our new breeding birds of the forest. It colonised Wales from the north of Britain arriving in mid Wales about 1980. It has now reached as far as the woodlands of south Brecon and Gwent. This dainty acrobatic finch is smaller than the Crossbill, being about the size of a Blue Tit but more slender in build. The plumage is green and lemon yellow, females and juveniles being paler and more streaked, while the male has a black bib and cap. Parties of Redpolls look similar in size and habits, especially when seen in winter flocks carousing through the crowns of alders or birches. If your opportunities to see detail are limited, any hint of green or yellow will tell you the birds are Siskins not Redpolls.

In springtime, from April till June, Siskins are busy building their tiny nests among the higher branches of a spruce or similar evergreen. The pair will fly in, landing near the top of their nest tree, which is often in a clump of similar firs, and drop to their nest below. As they fly above the trees they call persistently - a rather melancholic but high pitched 'pew' or 'pwee' note, delivered repeatedly. They nest socially and there may be several pairs nesting in the same part of the wood. The Siskin may well have prospered partly because it has adapted well to feeding at bird tables in winter. In the future the success of other species too may depend on garden food supplies put out for birds during the winter.

The Redpoll is a similar bird in many ways to the Siskin but its fortunes are in reverse at the present time, although the reasons for this are uncertain. In some places it is quite local although it is found throughout Wales. The Redpoll has two types of chosen habitat; stands of birch, alder or willow carr, usually on fairly low lying or damp ground, or conifer plantations where the trees are of low to medium height suitable for nesting in. The small black bib and crimson forehead are the best distinguishing features of this active little bird when the light is good enough for you to see them. Failing that, the high pitched trill uttered repeatedly as the birds bound overhead in animated fashion is the best distinction. Redpolls are especially noisy in June when they may be breeding sociably in considerable numbers.

The Coal Tit and Goldcrest are two of the commonest small birds of the conifers and their high pitched notes are regularly heard from the darkness of the fir canopy. The smaller bird with the squeaky voice is the Goldcrest. In springtime, provided the previous winter has not been too harsh and its numbers are not decimated, the woodlands may be alive with the calls of the Goldcrest, Britain's smallest bird. It is not an easy bird to see well as it flits between branches in the dark canopy of a spruce tree but sometimes, especially in winter, it will feed low in deciduous trees or visit gardens where its yellow crest draws attention to an otherwise tiny olive bird. The same-sized Firecrest is infinitely rarer and may no longer breed in Wales. A few pairs bred for more than a decade in the Wentwood forest in Gwent and others at Lake Vyrnwy in Montgomeryshire until a few years ago but they all seem to have gone. They are more likely to be found in mixed deciduous and conifer trees, especially perhaps those of exotic variety in parks or large gardens. Look for the double eyestripe (below and above the eye). The Firecrest has greener upperparts and paler underparts than the Goldcrest. Its song is stronger than the Goldcrest's and lacks the final flourish of the commoner bird.

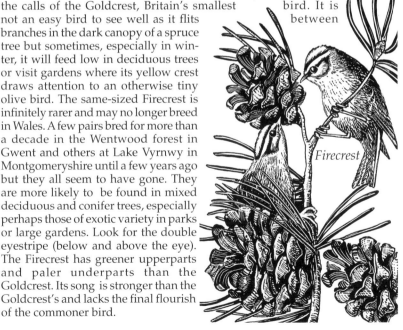

Firecrest

The Coal Tit is the commonest tit in the firwoods although there are good numbers of the more familiar Blue Tit and Great Tits. The Coal Tit is smaller than either of these two species and by comparison its colouring is more sombre. It has a black cap and bib and the white nape is totally diagnostic for this species. The song has the rhythm of a bicycle pump in action, more rapid and higher-pitched than that of the Great Tit which is rather similar. The Coal Tit usually nests in a hole at ground level among the roots of trees, an extraordinary position which may be related to the lack of other nest-hole sites in the conifer woods.

There are many other small birds which have greatly benefited from the proliferation of plantations, especially when the trees are small and there is a plentiful tangle of shrubs and brambles. Among the chief beneficiaries must be the Willow Warbler, one of the most abundant warblers which is especially common in Wales. It needs more open habitat than its close cousin the Chiffchaff, or the Garden Warblers and Blackcaps which are found in this habitat. The Garden Warbler in particular finds the understory of plantations and the upland deciduous woods much to its liking.

The plantations also provide an important refuge for the Song Thrush, which is becoming increasingly scarce in our hedgerows and gardens. At dusk the song of this familiar bird can be heard from all over the plantation as it repeats each note two or three times (at least), a trade-mark of this tuneful songster. Wrens, Robins, Hedge Sparrows, Pipits, Whitethroats and Bullfinches all find the more open parts of the woodland a congenial home to raise their young. The Yellowhammer and Grasshopper Warbler are scarcer, but the latter species can sometimes be detected on still summer evenings by its prolonged reeling notes delivered from the cover of sapling firs and tall sedges.

Among the larger birds, Wood Pigeon and Jay have benefited most from the conifer forests while Sparrowhawks build flat nests of sticks in the tall pine and spruce trees and hunt for prey along the rides and woodland edge. These birds are all seen regularly in the forest but there are rarer ones. These include two crepuscular species most easily seen at dusk or dawn.

The Woodcock is quite scarce in many parts of Wales during the spring and summer. It is found mainly in Gwent, parts of Montgomery and Breconshire (such as the Crychan Forest), more widely in Clwyd, and on RSPB reserves at Elan Valley, Ynys Hir, Lake Vyrnwy and at Rhandirmwyn in Carmarthenshire. It may be commoner than supposed since discovering one usually entails a spring evening committed to the darkness of the forest, or at least until the light has faded. Unusually, for a wader, it inhabits woodland - evergreen, deciduous or mixed - where it requires damp ground with ditches in which it can probe for invertebrates with its long beak. Visit a likely wood towards evening and wait, and look, and listen. You may first detect the presence of the male by the croaky voice calling 'chisick, chisick,' at several second intervals as he flies round the wood. The Woodcock comes into view just above the trees, slowly circling his territory. He passes your way perhaps two or three times more in the next ten minutes, and then disappears for the last time into the night. The light fails slowly and then there is silence. The Woodcock may appear the next evening, but there is no guarantee, particularly if the weather is poor. This courtship flight or 'roding', as it is called, may be observed at any time from March until June.

The Nightjar has something in common with the Woodcock. The courtship display is at dusk and both species are almost completely camouflaged against a background of leaves as they incubate their eggs. The Woodcock makes a scrape under a tree but the Nightjar scarcely bothers even with that and lays its two eggs in a slight depression on bare ground. The Woodcock is a wader like a very large plump Snipe, while the Nightjar hunts for moths, catching them in its wide gape which reminds one very much of a Swift. At dusk the Nightjar looks rather like a hawk as it twists and turns over the heath or cleared woodland. It lands lengthways on a horizontal branch of a tree or settles on a stump to pour forth its amazing reeling song, which has been likened to the sound made by a two-stroke motorbike. Sometimes it is detected by the 'whit whit' calls which penetrate the dark on calm evenings.

The Nightjar declined seriously over several decades, but in the past 10 years it has made a surprising and welcome recovery in Wales, probably due to the warmer summers; but this is not a species which keeps to warm low lying ground. On the contrary, most of Wales' Nightjars now breed in the uplands above 800 feet (240 metres). It has taken full advantage of the clear-felled areas of the forestry plantations and in 1992 a survey recorded a total of 193-197 'churring' males.

The woodlands of the Wye Valley in Gwent, parts of Breconshire, Clwyd and Glamorgan, are among the best parts of Wales to look for Nightjars. My own county of Ceredigion is not a good area, but this summer I decided upon a little exploration. I chose an area of upland pine forest which included young trees and scrub, ideal for nesting cover, as well as open space for feeding. There was a scattering of bushes and a few dead trees from which the Nightjar could deliver his long oration. I had waited in vain for a still, balmy evening but

Nightjar

there had been too much wet weather of late to give cause for optimism. I drove to the area and sat patiently. The evening was quite chill, and there was a disturbingly strong breeze, which I imagined was enough to dampen the ardour of most male Nightjars. At 10.15 p.m. I was delighted to hear my first Ceredigion Nightjar. His reeling lasted for about two minutes, after which there was a pause. A few minutes later he reeled again some 200 metres away and then silence fell but frustratingly the bird remained out of sight. As twilight dissolved into pitch black, I reluctantly turned for home. The following three weeks were very wet and cold but my next visit in July produced a repeat performance. This time the male bird did put in a brief appearance as he flew from the top of a sapling larch and vanished like a spirit into the dusk. I shall try again next year and hope that the Nightjar(s) - I am not totally convinced there was a pair- have more luck with the weather!

Among the most exciting species to be seen at the interface of moorland and forest is the Black Grouse. It returned in numbers to Wales after the new forests were planted following the Second World War, but in the past two decades it has gone into decline again as the forests have matured. It feeds on stubble, grain and grubs on farmland, on heather and bilberry and the shoots of sapling firs. In *In Search of Birds in Mid Wales* we wrote of hearing the males at their lek at Anglers Retreat in Ceredigion. This site is now overgrown as are

many others, rendering them totally unsuitable for this species. I also suspect that the Goshawk may take its toll, a burden which may be too much in areas where the bird is already scarce. Near Strata Florida, an area from where Black Grouse have not been reported for a long time, we recently found the feathers of a Blackcock dispatched by a predator, which may well have been a Goshawk. It is likely though that inadequate food sources arising from degradation of habitat rather than predation is the chief problem. We have seen single male birds at one or two places in the north of the county, but each year they become more difficult to find. Most leks in Wales consist of one male only. There are probably less than 300 birds left in Wales and the decline of this species is so rapid that there is a serious risk that it will become extinct in Wales within the next decade. The RSPB and other conservation organisations are currently making the survival of this fine species their top priority and the latest

Long-eared Owl

surveys indicate that they may be having some success.

There may still be one or two leks in the mid-Cambrians but most Welsh Black Grouse are to be found in the north of Wales, particularly in Meirionnydd, Clwyd and north Montgomeryshire. The male bird looks very distinguished in his blackish-blue plumage, thin white wing bars showing in flight, red wattle over the eye and lyre shaped tail. To hear him and his rivals, if there are any, bubbling and crooning at the lek requires an early start. An arrival on site at 6.00 a.m. between April and early June is strongly recommended.

At 'heart' the elusive Merlin is a species which is more at home on the moors than in the conifer forests and seems to have adapted to the fringes of them out of necessity rather than choice. There are, though, one or two benefits: the habitat produces a supply of food in the form of Chaffinches and other passerines, and Merlins' nests in thickly planted spruce are probably safer from predators and human disturbance than are those in moorland heather. This species is discussed with other birds of prey in Chapter Eight. These include the Honey Buzzard and Goshawk, which share an affinity for the coniferous forest.

The Short-eared Owl and Long-eared Owl also have their connections with the conifers though their habits and preferred haunts are quite different from each other. Both are rare in Wales as breeding birds. The Short-eared Owl, a species typical of heather and grass moorland, often hunts in the daytime over open country. It could, at first glance, somewhat remind the observer of a small headless harrier or even a Buzzard were it not for the owl-like 'flapping flight' and rounded wings. Usually it is not too difficult to see as it leisurely quarters the ground looking for voles. Occasionally a pair or two breed on a moor and then disappear after a year or two when the supply of voles dries up. The number of pairs in Wales rarely exceeds ten in any one year and this owl could turn up almost anywhere on moor, coastal marsh or on young conifer plantations (winter is the easiest time to see it, usually at coastal sites). In recent years a few pairs bred for a while at Trawsfynydd while at another time several bred on moors near Llanbrynmair. The Pembrokeshire islands of Skomer and Ramsey, alone in Wales, can claim regular breeding records.

The more nocturnal Long-eared Owl is a woodland species which keeps closer to cover in the daytime than the Short-eared Owl but comes out to hunt at night. It often breeds in small stands of conifer but is probably most easily detected by its dull triple hoot, heard usually in late winter or early spring. In the colder months this owl can sometimes be discovered at roosts (perhaps only a single bird) which are sometimes in marshy locations and in a thick cover of trees. The sleeping bird is almost invisible when perched in a protective tangle of bare oak, willow and hawthorn branches. 'Spotting the owl' is always a popular event if one should winter in a nature reserve, but if disturbed the owl may leave the site and go where it can find some peace! The Long-eared Owl is sometimes given away by the actions of mobbing passerines. I once discovered one sitting bolt upright against the bole of a birch tree, drawn to my attention by the agitation of a party of Long-tailed Tits. Occasionally pairs are located in different parts of Wales, and it is probably often overlooked. Interestingly, an observer using a tape-lure in Breconshire (the male bird responds to what he thinks is the voice of a rival) obtained responses at six sites in the county. This is remarkable for a species which passes almost unrecorded in most counties from one year to the next and suggests there may be many more Long-eared Owls in Wales than we have supposed.

CONIFEROUS FOREST - species to look for:

Coal Tit

Goldcrest	Grasshopper Warbler
Coal Tit	Goshawk
Wood Pigeon	Crossbill
Jay	Nightjar
Sparrowhawk	Woodcock
Buzzard	Black Grouse
TreePipit	Merlin
Siskin	Short-eared Owl
Redpoll	Long-eared Owl

Notes: (1) The commonest species are likely to be Chaffinch, Coal Tit, Goldcrest, Robin, Wren, Song Thush, Wood Pigeon, Blue and Great Tits, together with summer migrants such as Chiffchaff, Willow Warbler, Blackcap and Garden Warbler; in other words, many of the most abundant species described in Chapter One.

(2) Siskin and Redpoll are locally common, Crossbill and Goshawk are worth looking for in suitable woods, some of the other specialist species in the right-hand column are distinctly scarce or even rare.

(3) The species are listed from top to bottom and from left to right in order of rarity, from the abundant Goldcrest to the rare Long-eared Owl.

CHAPTER FIVE
BIRDS OF THE RIVERSIDE

Fast flowing rivers are an integral part of upland scenery and Wales of course, has an abundance of both. The nascent streams, fed by saturated peat and slippery rocks dripping with water, spring to life at their sources high on the moors and mountains. Tumbling down hillsides or leaping over rock faces, they come almost alive as they ripple and gurgle, swirl and foam, or squeeze through narrow ravines in their headlong rush towards the sea.

Swollen by tributaries, the sparkling river widens and slackens in pace as the gradient declines on the tortuous route to the valley floor. There its winding course is often marked by overhanging ash or alder trees, or oaks covered in lichens from the action of rain and river spray. Where the meandering river cuts deeply into the soil, banks of sand and clay are formed, or patches of gravel and shingle are strewn along the water's edge, to be submerged only in times of flood. Wales has many beautiful rivers whose moods and character change during their journeys to the sea but invariably they start in haste and finish at a more leisurely pace. The Severn, Wye, Tywi, Teifi and Dee are among the largest rivers but there are others too numerous to mention, pouring their waters into the Severn estuary or the Irish Sea.

The different sections of the river are favoured by different species of riverside birds. Rushing streams and water-falls with steep banks of ivy, or stone bridges and other artefacts, attract Grey Wagtails and Dippers. Broader, slower stretches of river provide ideal habitat for Sand Martins and

Dipper

73

Kingfishers which bore their nest holes deep into the river banks. Sandpipers may choose either, provided there is shingle or other level areas with vegetation for nesting. There are some interesting and very characteristic species then, which can be enjoyed in a relaxed fashion as you stroll or take a picnic on the banks of a river almost anywhere in Wales.

The Dipper is one of the most typical of the river birds, although it has become scarcer over the past few years. The size of a thrush, but more the shape of a Wren, the plump, rotund Dipper inhabits the river all year round, even in the depths of winter. Often it first attracts attention by its crisp 'zit' call and white flash on its throat as it flies with rapid wing beats to land on a midstream rock. It hunts for its invertebrate prey consisting mainly of small crustacea, walking adeptly on the bed of the stream in its search for food. The nest is domed, rather like that of a large Wren and is placed on the ledge of a stone bridge, or hidden among ivy in rocks where it is often sprayed by a gushing waterfall.

In Britain it is probably more common in Wales than most places, while abroad it inhabits the mountains of Europe and ranges from places as far apart as the Himalayas and the American Rockies. Recent declines in Wales are thought to be connected with pollution or increased water acidity. In my own district, formerly I could expect to see them in several places on the Rheidol and Ystwyth from where they seem to have vanished, temporarily at least. A brief search on the Severn however, yielded sightings of four birds along a short stretch near Llanidloes in only about two hours, while Dippers can readily be seen next to the Taff Trail close to Cardiff. Research is being carried out which should be helpful in suggesting the conditions required to restore our Dippers to their former numbers.

The typical species you are most likely to see by the riverside is the Grey Wagtail although the Pied Wagtail, less fussy about its habitat requirements, is observed at least as often. At other times of the year the Grey Wagtail strays into gardens and various other habitats, but in the breeding season it keeps close to the river, especially where the stream flows through wooded country. The name 'Grey Wagtail' greatly underestimates this lovely bird which has grey-blue upperparts and primrose underneath turning to vivid yellow on the under-tail coverts. It habitually flies from rock to rock, wagging its tail vigorously, lunging upwards, almost hovering over the stream as it snaps at passing insects. Usually the nest is placed on a ledge under a bridge or in the ivied rootlets of trees on the banks of the river, but sometimes it chooses a platform or crevice on a rockface or bank some distance from water. Watch the pair collect beakfuls of insects and fly a succession of sorties along the river to feed a nestful of hungry mouths.

In April the Common Sandpiper returns from Africa to nest on the shores of our rivers and upland lakes, although it is commoner further north, especially in the Highlands of Scotland. The Sandpiper is a charming excitable bird during its courtship display, as it bobs up and down, chases its mate along the shingle banks, or flies over the river on stiff bowed wings. The female lays its eggs in a depression which may be in a tuft of grass or other herbage, or it may be on the bare shingle. On some barren upland pools, this species and the Pied Wagtail may be the only birds breeding on an otherwise desolate shore.

The sawbills, or to be more specific the Goosander and Red-breasted Merganser, are also birds of the river, having colonised Wales from the north. Both species first bred in Scotland in the mid 19th century and the first Welsh birds were presumably either from there or possibly from Cumbria. Both have long serrated bills especially adapted for holding slippery trout and other fish. The larger Goosander first bred in Wales in Montgomeryshire in 1970 and its progress throughout most of Wales since then has been meteoric. It is found commonly on rivers like the Teifi, the Tywi, Wye, Severn, and Dyfi. The easiest way though to see Goosanders is probably to look on the larger lakes such as Vyrnwy or Tal-y-Llyn in winter, when the cream and black males are very conspicuous. The Goosander usually lays it eggs in a hole in a tree or riverbank. The only eggs I have seen personally were in an oak tree, but its two eggs were being incubated by a Tawny Owl which had presumably ousted it from its nesting hole and had even laid one egg of its own! The eggs hatched but I cannot imagine the young Goosanders would have survived long on a diet of mice and shrews!

The Red-breasted Merganser has bred in Wales for a longer period and reached the Ceredigion shores of the Dyfi in 1969. There its southward momentum appeared to be arrested until 1985 when we proved breeding on the Ystwyth. (See *In Search of Birds in Mid Wales* published 1988). Unlike the Goosander, the Merganser makes its nest in tall herbage rather than in a hole in a tree or rock, but the eggs, unlike those of most other ducks, are laid at the end of a long runnel through the riparian vegetation. The Merganser is found mostly near estuaries or rivers close to the sea in north-west Wales and its expansion has not progressed materially towards south Wales. In 1995 though, the first pair did breed in Pembrokeshire at the Gann. On the Ystwyth we have the Goosander and Red-breasted Merganser overlapping in their breeding territories along the river.

Common Sandpiper

The males of the two larger sawbill species (the rare Smew is also a sawbill) are quite easy to tell apart. The male Goosander is larger and has much more white or creamy plumage. In flight he shows white shoulder patches like the Shelduck. The females are harder to distinguish: both have grey bodies and reddish brown heads. Look at the heads closely. The Goosander has a smooth rounded contour while the smaller Merganser has a double crest. In some ways they are rather strange looking birds and do not share the plump 'cuddly toy' image of most duck species. Instead, the slender form, cold red eye and long 'toothy' beak, may convey a hint of reptilian ancestry in the make up of these birds.

The species we have considered so far are all characteristic of highland Britain. Some of those which follow are more numerous in the lowland east where the rivers wind slowly along broad fertile valleys. The Kingfisher is a bird more typical of sluggish rivers with ample banks of sand eroded by the meandering movements of the water currents. In Wales it is most numerous in the south and east on rivers such as the Tywi and Severn. For me, in west Wales, it is an unusual treat to see the luminescent blue flash of a Kingfisher race under a river bridge, or dart from an overhanging branch where it has been fishing on a still pool isolated from the main thrust of the river. A few can be seen close to the Dyfi and Mawddach in west Wales, and on rivers such as the Rheidol and Ystwyth they may be encountered on the same stretches of river as the Dipper. The same general comments may be made about the distribution of the Sand Martin which, like the Kingfisher, bores its nest hole a metre long inside a sandbank. It is most common in south Carmarthenshire and Gwent. The Sand Martin nests in colonies, and sizeable communities of them can be found at places like Glasbury on Wye or at points along the River Severn. In more upland districts where suitable sites are at a premium, a few pairs are found on even tiny upland streams, or in the low banks of fast flowing rivers.

Often in late summer the Sand Martins fly among swarms of Swallows and House Martins in pursuit of airborne insects over villages, rivers and reservoirs. The House Martin has a distinctive white rump and inky blue-black upperparts which makes it easy to separate the two species of martin. The Sand Martin has a sandy-brown back and wings and a dark band across the upper chest. Both martins lack the long tail streamers and maroon-red throat of the Swallow.

The Heron is a typical waterside species though it is just as likely to be seen standing motionless at the margins of a shallow lake as by a river, waiting to stab any unwary fish that comes within striking range of its formidable dagger-like beak. This weapon, unleashed with muscular strength and leverage, is also a danger to mice, rats, frogs and small fledgling Coots or Moorhens.

Herons in Wales frequently nest in large trees close to rivers, constructing their fragile nests often at considerable height above the ground. Like Rooks they nest in colonies, usually laying their pale blue eggs in late February or early in March. Despite being one of our most familiar and easily recognised birds, they are quite local in distribution. Unless we count the Bittern, they were until recently the only heron species regularly to be seen in Britain. This is just changing however, as the number of Little Egrets visiting Wales grows year by year. Probably 50 birds or more are counted in most years, especially in winter in South Wales. I was surprised to see one myself on the River Ystwyth as long ago as Boxing Day 1989. This species, attracted no doubt by warmer weather in Britain during the past two decades, is already nesting at two sites in Dorset. In 1998 a pair made a nest in Pembrokeshire and although no eggs were laid we can confidently expect this unlikely addition to our avifauna to go from strength to strength. We may see more of other members of its family in future years such as the Night Heron,

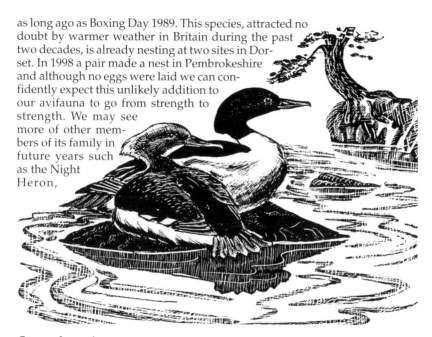

Goosander pair

Purple Heron and Great White Egret, species which are now met with rarely, mainly on South Wales lakes and marshes.

Perhaps the commonest bird in these habitats is the adaptable Mallard which seems to thrive in all but the most turbulent reaches of Welsh rivers. In its 'pure bred' form, the drake is easily distinguished by its bottle-green head below which there is a narrow white ring separating the neck from the rich purple-brown chest. Both sexes have a distinctive blue speculum on the rear edge of the wing. Feral birds, many of them hybrid specimens displaying various plumage types, can usually be seen around villages streams where they are sometimes fed by local people. At higher altitudes one or two Teal may replace the Mallard, but generally, except near the coast, there are few wildfowl of any kind on the rivers themselves except sawbills, Mallard and in some districts, Moorhens. The latter species is also found on the canals of eastern parts of Wales, but like other waterfowl it may often fall victim to the American mink, which has prospered on many waterways following its escape (or illegal release) from mink farms in Britain.

We have almost concluded our birds closely associated with Welsh rivers but there is one exciting species which started to breed beside them in the 1980s and is doing well. The Little Ringed Plover first came to southern England in the 1930s and after the Second World War capitalised on the sand and shingle quarries created to service the post war building industry. I was lucky enough to find one of the first breeding Little Ringed Plovers in Warwickshire in 1959. This dainty plover brought a new ornithological interest to sand and gravel workings and became one of the most popular species among birdwatchers in places like the Midlands. It is smaller than the Ringed Plover with dark beak and greenish legs in place of the flesh-orange colours of the larger bird. It has a different head pattern and a noticeably pale eye ring, but lacks the wing bar which is so prominent in the Ringed Plover.

The interesting thing about our Welsh birds is that they are colonising the shingle banks of rivers, not sand and gravel workings. This habitat is used on the continent but hitherto not in Britain. They lay their four eggs in a scrape either on bare ground or on shingle. Crows are probably the main threat to the eggs, but Buzzards, Kestrels and even Herons will take fledgling birds. It is important therefore to avoid disturbing plovers at their nest sites, particularly since the restricted habitat of narrow river bank makes them more vulnerable.

The main rivers claiming Little Ringed Plovers are the Tywi Severn, Usk and Wye, where the switchback twists and turns of the rivers leave extensive banks of exposed pebbles, especially when the water level is low. In North Wales two or three pairs currently breed at the RSPB reserve near Conway and could spread to other locations on the north coast. Hopefully this delightful species will eventually colonise many suitable rivers throughout Wales. In western areas the Little Ringed Plover is still rare but in 1999 we located a pair in Ceredigion which bred successfully. After discovering the birds flying close to the river in April, we visited the spot again two weeks later. Some distance from the site we spotted the pair through our binoculars, feeding at the edge of the water, and waited to see what would happen next. After a few minutes one of the plovers walked slowly and deliberately for about 10 metres and stopped to look around before shuffling deep into the pebbles near the top of slope. The incubating bird was so well camouflaged on the shingle bank that it was difficult to spot her even with the aid of a telescope. A final visit to the site in June revealed four marbled youngsters running after their mother along the water's edge, sometimes freezing at some unknown signal from the parent bird until a perceived danger had past. Hopefully this young generation will survive to populate other rivers in the area in coming seasons.

There are one or two other interesting species to discuss, although strictly speaking they are birds of river valleys in Wales rather than the rivers themselves. The Yellow Wagtail is a scarce summer visitor to Wales. Its main concentration in Britain is in the east of England as far north as Yorkshire and then down to the home counties, where it favours wet marshy meadows and adjacent arable land. It either nests in these wet pastures or in fields of corn or barley. Mixed arable farming close to lush river meadows is in short supply in Wales but you can find it in some areas. Farmland near Glasbury on Wye, close to the border with Herefordshire, provides ideal habitat and there are several pairs in that district. There are more Yellow Wagtails close to the Severn and others breed sparsely on the Gwent Levels but this species is always likely to remain scarce in Wales for the same reason that the Grey Wagtail should remain common (namely, the availability of suitable habitat).

The male Yellow Wagtail is a striking vivid yellow bird when you see him with full light on his plumage. He perches lightly on the topmost twigs of a roadside hedge, balancing with up and down movements of the tail, or hovers delicately above ripening barley concealing the nesting site tucked into some clod of earth among the cereal crop. Pipit-like in build, but with a long tail, the Yellow Wagtail has a delicacy of movement exceeding even that of its dainty cousins, the Pied and Grey Wagtails. The female is paler in coloration, but is otherwise similar in appearance.

The Tree Sparrow has less claim to be a 'riverside bird' than the Yellow Wagtail but in Wales it is often indirectly associated with them. It is in the broad lower river valleys that the most fertile soils occur and in such conditions there is more likely to be greater diversity of agriculture. Harvested crops may leave a residue of stubble and seeds to provide winter feed for this increasingly uncommon bird. Currently the greatest numbers are reported from the Wye valley in Radnor and Brecon where winter flocks may number 50 birds or more. This modest but rather neat and appealing bird differs from the larger House Sparrow in that the cap is a uniform warm brown colour, and there is a chocolate 'comma' mark on the cheek. Unlike its near relative, both sexes are similar in appearance. The Tree Sparrow used to be a very common bird, nesting socially in holes in decaying oak or ash, or sometimes placing its untidy straw nest filled with feathers in the ivy creepers clinging to mature trees. It has been estimated that as much as 90% of the Tree Sparrow population has been lost in recent decades, and it is becoming relatively scarce even in its strongholds in the arable districts of England.

Some species may be seen commonly near rivers but not exclusively so. They include such species as the flycatchers, Marsh and Willow Tits, Redstarts and Lesser-spotted Woodpeckers. Many small land birds like the vicinity of water because steep banks and overhanging bushes afford protection for nests. There is also a good supply of airborne insects over water and grubs in rotting trees and other decaying vegetation. Other species occur on rivers, usually where they flow through lakes or marshes and these will be discussed in the following chapter.

THE RIVERSIDE - species to look for:

FAST FLOWING RIVERS	SLOW-FLOWING RIVERS
Pied Wagtail	Mallard
Grey Wagtail	Canada Goose
Dipper	Mute Swan
Common Sandpiper	Heron
Goosander	Little Egret
Red-breasted Merganser	Moorhen
	Little Ringed Plover
	Kingfisher

Notes: (1) There is no strict division between these two groups on Welsh rivers, but there is a definite preference by different species for one type of water-course or another. Those birds listed in the left-hand column are typical of highland Britain, and those on the right generally flourish in lowland districts.

(2) Species associated with lakes and marshes such as Dabchick, Teal and Coot may also be found on the rivers.

Pied Wagtail

CHAPTER SIX
LAKES, POOLS AND MARSHLAND

Although Wales has some beautiful lakes and reservoirs, its reputation for outstanding scenery does not rest with them. Valleys of deciduous woodlands, rivers and mountains lie at the heart of the Welsh countryside but lakes are on the whole less noteworthy compared, say, to Scotland's lochs or Cumbria's lakes. The same holds true from the birdwatching point of view. Generally speaking, most of Wales' fresh water lakes are too acidic and too deep to attract the wealth of species found in lowland waters, while the Scottish specialities, the divers, scoters, Goldeneye and Slavonian Grebe, do not breed so far south as Wales.

In many parts of Wales even the Moorhen is a relative rarity, while many upland waters can claim just a pair or two of Mallards or Coots in the spring and summer months. Even larger lakes and reservoirs such as Llyn Brianne support only a meagre range of breeding species - Coot, Mallard, and Goosander. Some lakes have colonies of Black-headed Gulls which announce their presence with a raucous cacophony of noise between April and July. Occasionally colonies of these gulls disappear for no apparent reason. The 200 plus pairs of Black-headed Gulls breeding on an island at Llyn Syfydrin in Mid Wales vanished quite quickly. When I last visited this formerly noisy place in May there were just a few Mallards and Coots and one lovely Black Tern hovering over the surface of the lake, dipping its beak into the water to feed

Teal

on surface insects. This upland lake was an unusual sight for this attractive migrant. More often they are seen over marshy pools and lakes at lower elevation, or over the sea, particularly in August and September. Species of birds which can be expected on such upland lakes are not only sparse but predictable, so we were pleasantly surprised early last May to find a pair of migrant Garganey floating on a small pool high in the hills between Tregaron and Abergwesyn. This is a scarce duck in Wales, and our pleasure was compounded by the presence of a pristine male Goosander resting on a low mossy bank close to the water!

Teal still breed, though in decreasing numbers, on some upland peat pools especially in Ceredigion but the majority of this species are found on lower ground at places like Cors Caron. The 1995 survey of 150 square kilometers of the Elenydd discovered only eight pairs, of which just one was proved to breed. They used to be found near to the mountain road between Rhayader and Aberystwyth, but are rarely seen there now. Sometimes the Teal may incubate its eggs well away from extensive water. Some years ago we inadvertently flushed a female from a nest containing nine creamy white eggs, well concealed in reeds and sedges beside a twisting moorland brook, a good distance from the nearest pool or lake.

Some of the shallower, lower lying lakes surrounded by shoreline vegetation are rather better for birds. This is particularly true if they lie over basic limestone rocks which support a greater variety of plant life, and in turn, invertebrate creatures on which both fish and birds can feed. Llangorse Lake in Breconshire is the outstanding example and this lake supports good numbers of wildfowl and waders at all times of the year. I remember visiting for the first time (in winter) and immediately observing a flight of seventeen Shoveler over the lake. I wondered where else (with the exception of Anglesey), this number of what is a rather an uncommon duck, could be seen inland in Wales. It lies in a green fertile valley near Talgarth, and has breeding Snipe and Redshank, Water Rail and Reed Warbler and of course, Great Crested Grebe. This bird is by no means a common breeding species in Wales, but the lake holds up to a half a dozen pairs. Tal-y-Bont reservoir, in the same county, sometimes supports two or three pairs.

The long-necked Great Crested Grebe is an elegant bird in summer when the courting pair pirouette, shake their heads, and touch beaks in the middle of their breeding lakes. Both birds sport impressive crests, and the cold winter plumage is replaced by warm chestnut-brown on head and back. In west Wales the Great Crested Grebe is particularly scarce. In my own area of Ceredigion there are just two or three pairs breeding in the whole of the county, for instance at Llyn Eiddwen on the Mynydd Bach. Further north, in Meirionnydd they occur on Tal y Llyn, a beautiful setting amongst the hills of Cader Idris and at Lake Bala (Llyn Tegid), the largest natural lake in Wales. There are more of these grebes in eastern Wales at places like Lake Vyrnwy and Llyn Mawr in Montgomery, scattered pairs in the south, and many more on Anglesey. Even the chubby, chestnut-cheeked Little Grebe or Dabchick, which can skulk invisibly in waterside vegetation, is quite scarce in Wales. Though less conspicuous than the larger species, it often draws attention to itself by diving repeatedly in the open water as it searches for small fish. Like other grebes, the

Dabchick makes a soggy nest of rotting vegetation which appears almost to float on the water but which in reality is attached to small sunken branches, submerged reeds, or similar anchorage.

In general, Wales is not especially noted for its breeding wildfowl. There are good numbers of Goosanders and Red-breasted Mergansers which we dealt with in an earlier chapter, but such species as Gadwall, Pochard, Shoveler and the rare Garganey are seldom met with in summer in many parts of Wales. Anglesey is an important exception to the rule. Our northern island county is for the most part quite flat. On first acquaintance it seems like an island which should have mountains, since it has rocky outcrops with areas of gorse, and fields bordered with stone walls which remind you of upland areas. Interspersed in this landscape are numerous lakes, set mostly in limestone rocks and rich in bird life. Among these are the Valley lakes, Llyn Alaw and Llyn Llywenan. The last named lake is noted as the place where the first ever breeding of the Black-necked Grebe occurred in Britain in 1898, and continued at this site for the next 30 years. This rare grebe is still encountered on passage in Wales and could breed again in the future, as it still does on a few waters in central Scotland.

Take a trip to Anglesey and start with the lakes close to the R.A.F. training base at Valley. If you don't mind the roar of military aircraft taking off or flying over the base, you will be rewarded with several species of duck on the RSPB managed lakes. Male Tufted Ducks in their pied plumage and the red-headed male Pochard will be found diving in the turbulence of the wind blown waters. The Gadwall and Shoveler may also be seen on these lakes, although the latter species is much commoner there in the winter months. The male Shoveler, with his long spatulate bill, blue-black head and orange-brown and white body plumage, is a very striking bird. His more sombre looking partner is best told from the female Mallard by her shorter neck and characteristic beak. The profile of the Gadwall is similar to the Mallard, and the duck of the two species are not always easy to tell apart. Both she and the male have a white speculum on the wing instead of a blue one as in the case of the Mallard. The drake Gadwall is a distinctly greyish bird with black under tail coverts. The black-bordered white rear wing bar of both duck and drake are quite distinctive in flight. One of the most colourful ducks is the Ruddy Duck, the male of which 'has blue bill, rich brown and white plumage and stiff up-ended tail. This immigrant from North America, now established as a British bird, is found at more than one site on Anglesey.

Two or three days spent on the island should enable you to see as many as eight or nine species of breeding duck (the six mentioned above plus Teal, Shelduck and Red-breasted Merganser). In winter, you should add several more to your list: Wigeon, Pintail, and Goldeneye are usually on the lakes or estuaries, Common Scoter and more rarely, Scaup and Velvet Scoters, are more exclusively to be found on the sea. If you are very lucky, the scarce Long-tailed Duck may turn up on a mere or in a shallow bay.

Several of the Anglesey lakes are bordered by broad margins of phragmites, the tall, slim reed favoured for nesting by both Marsh Harrier and Bittern. Both of these species have bred on Anglesey since the Second World War, and should either return to breed in Wales, some of these reed beds are possible candidates to welcome the first pairs. Reed Warblers, which are still scarce in Wales and confined to this kind of habitat, already pour out their grating notes from them between May and August. The secretive Water Rail, far less easily observed even than the Reed Warbler, squeals and grunts its pig-like notes at night or in the dawn mists as its narrow body slips between the stems. On more open wetland sites, Anglesey can also claim the distinction of breeding records for two rare waders, the Ruff and Black-tailed Godwit, two species more or less confined as regular breeding birds to the fens of East Anglia so far as Britain is concerned.

Elsewhere in Wales, extensive stretches of fresh water surrounded by richly vegetated margins are limited, although there are some smaller lakes in eastern Wales which have breeding wildfowl. Llyn Coed y Dinas and Llyn Newydd in Montgomery, for example, support one or two pairs of Tufted Duck. This species used to be quite scarce in Britain but its expansion in the West Midlands, like that of the Great Crested Grebe, accompanied the post-war excavation of new gravel pits to meet the building boom of those years. There seems every chance that the Tufted Duck, which seems to occur ever more widely at all seasons in parts of Wales where it was formerly infrequent, will expand its range as a breeding species. Already it is a familiar bird the year round on some lakes near Cardiff and elsewhere in South Wales. There are one or two places near the mouth of the Tywi and further north in Carmarthenshire which attract breeding duck such as the Tufted Duck and Pochard.

In various parts of west Wales in particular, there are some residual lowland heaths and mire carpeted with heather and grassy patches rich in unusual plant life. Some of the larger ones are under the auspices or ownership of conservation organisations, places like Dowrog Common near St David's in Pembrokeshire. The main interest is usu-
ally botanical while pipits and larks predominate among the bird life. Whinchats and Reed Buntings are normally present in the damper areas. Grasshopper Warblers and Stonechats may feature, while patches of willow carr frequently attract Lesser Redpolls and Willow Tits. The larger pools may have breeding Teal, Dabchicks, Water Rail or Black-headed Gulls. Winter offers more promise for visiting birders and at this time of the year interesting birds of prey may be encountered. Ceredigion has two extensive tracts of peat bogland of this nature, the vast Cors Fochno near Borth and the even more extensive Cors Caron. Large tracts of both bogs are National Nature Reserves managed by the Countryside Council for Wales. There are Curlew, Snipe and a few Redshank still breeding in and around these marshes, but Red

Cetti's Warbler

and Black Grouse and Dunlin, all of which used to nest on Cors Caron in the 1930s, have long disappeared. It is worth keeping alert, especially when the light is fading, for the distinctive 'whiplash' calls of a Spotted Crake in this type of marshy habitat. This secretive bird, the size of a thrush with a corn-yellow beak, is occasionally heard but rarely seen. It is spasmodic and irregular in Britain, but there are signs of an increase in number and frequency of breeding records lately. It has bred in Wales in the past and could well do so again.

The Teifi marshes, together with such places as Oxwich on Gower and Kenfig Pool in Glamorgan, have thick beds of the reeds which attract such species as Reed Warblers and Water Rail, and potentially Marsh Harriers and Bitterns. Included among these more exciting possibilities is the Bearded Tit, which has bred occasionally in Wales in recent years. This species used to be confined to East Anglia but has greatly increased after near extinction during the cold winter of 1963. It now breeds in Humberside, Lancashire and Dorset and could well establish itself in Wales. Both the Bearded Tit and the Bittern are 'prize species' which are sometimes encountered in winter, maybe driven westwards by the harsher weather conditions in the east. The prevailing reedland habitat at these sites is so limited in Wales that its importance for those species which depend upon it can hardly be exaggerated.

The Teifi marshes are owned by The Wildlife Trust West Wales. This reserve, straddling the border between Ceredigion and Pembrokeshire is an interesting place to visit at any season but especially in spring and summer. Take a walk along one of the main tracks and listen and look for the many species of warbler which breed on the reserve. 'Listen' is the operative word, since these 'little brown birds' keep to cover and are often reluctant to come into the open. Two of these warblers of the marsh, the Reed and Sedge Warbler, sound quite similar and it will take the inexperienced birder some practice to separate them. Both of them are enthusiastic mimics of the songs of other birds. The Reed has

a less varied range of notes than the more vibrant and commoner Sedge. Both, it must be said, are totally lacking in musical talent when compared with other members of the family. When they poke their heads above the reeds however, identification is simple. The Sedge Warbler is more striated and has a clear eye-stripe compared with the more uniform plumage of the Reed Warbler. The nest of the colonially breeding Reed Warbler is one of the most beautiful constructs among British birds, a nest of fine grasses, moss and wool, interwoven between three or four tall reed stems over water.

There are some much rarer warblers of the marsh. The Savi's Warbler has been heard in Welsh

Sedge Warbler

84

reedbeds while the Marsh Warbler has bred here in the past. The Aquatic Warbler, a rarity from Eastern Europe, has been observed or caught in mist nets on a number of occasions. There appears to be no limit to the gems which may be concealed within the depths of a reed bed! There is one former vagrant however, which now breeds annually in Wales.

As you walk along the track at the Teifi Marshes, where brambles and willow carr tangled with reeds grow on both sides of the path, stop and listen attentively. Suddenly the silence is broken by a loud explosive note, 'chewi chewi chewi'. It has the same sudden, loud and explosive quality of the wren, though the song is very different. The sound stops as abruptly as it starts, but where is the bird? You detect a slight rustle among the tangle of brambles and reeds, but otherwise there is no sound or movement. Just when you have given up, there is another burst of song 20 metres away, and then silence again. The Cetti's Warbler is just about invisible, but if you should catch a glimpse of one in reasonable light the rufous tones of the plumage are very noticeable. It is very much the archetypal small brown bird, though slightly more robust in build than the Sedge and Reed Warblers.

The Cetti's Warbler needs to be robust. It is one of only two resident warblers which brave the winter here although as we have seen, a number of Blackcaps and Chiffchaffs also spend the coldest months in Britain. If you visit Greece or Spain in summer, you may often hear the Cetti's Warbler singing from almost insignificant damp or reedy ditches. In Britain it first nested in Kent in 1964 and spread westwards. The cold Kent winters, however, proved too much for the Cetti's Warbler, and the main concentration of the species is now in Hampshire and counties to the south and west. Wales holds maybe as many as 30 or 40 pairs but the population is very vulnerable. In 1995, an estimated 23 pairs at Teifi had slumped to just a few pairs by 1998. You will still hear them at Oxwich and in one or two places in Pembrokeshire and south Carmarthen not far from Llanelli.

The other resident British warbler is the Dartford Warbler which inhabits dry Gorse and heather-clad commons, not marshes. This attractive little bird with purplish-grey and maroon plumage and long tail, was recorded nesting for the first time in Wales in 1998. In southern England it is flourishing after nearly becoming extinct there in the sixties. There is plenty of suitable habitat for it here, so it could well be one of the first species to colonise South Wales in the new millennium!

The Grasshopper Warbler inhabits marshes, although this species may be found in drier areas with deep tangled grasses on plantations. It nests in long herbage and does not require the wet reedy situations beloved by Sedge and Reed Warblers. Last season I heard four males on Cors Caron and there may have been more. The reeling song is actually like the sound of a grasshopper as the name implies and may continue for 30 seconds, a minute, or even longer. Listen carefully in suitable habitat. The pitch of the reeling may be hard to pick up, but when you hear it this species is difficult to mistake for any other bird (except possibly the very rare Savi's Warbler). The other small bird associated with reeds is the widespread Reed Bunting, which may occur in quite small patches of reed on damp farmland fields. This species has a simple song but fortunately is apt to sit more conspicuously on top of bushes or reeds than most of the warblers do. The breeding male has a black head and white collar, and the head pattern of both sexes is quite distinctive at all seasons.

Possibly the most popular type of habitats for birdwatching are the muddy margins, scrapes and nutrient-rich stagnant marsh surrounding freshwater pools. Often the best of these are located at nature reserves, in most cases not far from the sea.

Snipe

In this setting groups of birders can be seen eagerly scanning every square metre from the concealment of a reserve hide. Wildfowl, Water Rail, Kingfishers, gulls and grebes, may all be the object of excitement and anticipation. In this atmosphere a sense of cameraderie usually prevails, information is readily shared, and telescopic close-ups are enjoyed by experts and novices alike. Of all the different groups of birds, the waders are probably most often the target of interest, particularly the more unusual ones occurring fleetingly on passage. In late summer, there may be a Greenshank, a Spotted Redshank, or even a Wood Sandpiper. September may bring Little Stint or Curlew Sandpiper. There are absolutely no limits to the bounds of possibility - in theory at least - and news of real rarities travels fast. White-rumped Sandpiper and Marsh Sandpiper from north Europe or Asia, and accidentals from America like the Terek Sandpiper or Wilson's Phalarope are all possible. All of these and many others have been seen in Wales. Small wonder so many are smitten by the twitching bug in these days of easier transport and telephone bird lines!

Returning to those species which are a little more typical of Welsh lakes and marshes, what else can reasonably be expected? The tall Grey Heron is, of course, a very familiar sight standing patiently in water, seemingly indolent but deceptively alert, or tip-toeing gingerly through the reeds, searching for frogs or other prey. At a distance in flight the Heron may be mistaken for a large raptor, but its bowed-wing silhouette, observed from a head-on perspective, is characteristic of the heron family. The same is true of the folded neck which distinguishes herons and egrets in flight from similar sized Storks, Cranes, and Spoonbills, which all fly with their necks extended. The White and Black Storks and the Crane are rare vagrants in Wales but you may be lucky enough to come across a Little Egret or even a Spoonbill. Its snowy white plumage, black bill and legs and bright yellow feet make the Egret almost impossible to confuse with any other wild bird you are likely to see regularly in Britain. Like the Little Egret, the Spoonbill is also white but much larger, and flies with rapid beats of its short wings in contrast to the more leisurely heron-like flight of the other bird.

Last summer I was talking to a retired man who lived in a bungalow ideally situated for birdwatching close to the Dyfi estuary in Mid Wales. He told me of the thousands of Wigeon and much fewer numbers of geese he had counted as they flew to browse on the salt marshes further up river. He also mentioned that two Little Egrets and a Spoonbill had spent several months on the marshes during the summer. The Egret used to be a rare vagrant but now, probably as a result of a succession of warmer summers and mild winters, it is to be seen increasingly in estuaries and rivers in several places in southern England and in South Wales. On a far less dramatic scale the same might be said of the Spoonbill, which is now recorded on several dates in most years in Wales. We may regard this as one of the positive effects of global warming due to the greenhouse effect. There will be other less desirable ones - flooding and unpredictable weather. Nevertheless, we are witnessing more exotic visitors, including Hoopoe, Golden Oriole and Purple Herons. It is possible, in years to come, that we shall find them all in Wales, although it is unlikely that the weather will be good enough for the hard-pressed farming community to turn from the sheep economy to grapes!

The increase in sightings of Marsh Harriers though, probably has a different cause. Like the Bittern and Bearded Tit it has large populations in Holland, whilst in England it is especially familiar in East Anglia. There it tottered for eighty years between extreme rarity and extinction. At the turn of the century it disappeared entirely, returning to the Broads at Hickling in 1911. Until the seventies there were only a few pairs, either there or on the Suffolk coast. For some years these three or four pairs, declining to only one at its low point in the mid-seventies, were the only ones nesting in Britain. Many ornithologists believed the species could never become anything more than a rarity because it required extensive beds of phagmites reed. To everyone's surprise, the Marsh Harrier overthrew the confines of its reedland home and started nesting in cornfields. A third of East Anglian Marsh Harriers now breed in fields of wheat and barley. At the same time, they have expanded to places in northern England, Kent and Somerset.

In the 19th century, before persecution and drainage of their wetland led them to extinction, they were found in many areas of the British Isles, as far west as Ireland. At one time they bred at various places in Wales including Crymlyn Bog near Swansea. Increasing numbers of Marsh Harriers are now seen in Wales though we are still waiting for the first nesting attempt since 1974. They occur at various times of the year, usually on spring or autumn passage, but sometimes deep into winter or in summer.

Last June I was delighted, but not entirely surprised, to see a Marsh Harrier quartering an area of suitable territory in Wales. At first I thought my bird was a Buzzard. The two species are similar in size and wing-span, and after all I had looked hopefully at scores of similar birds flying low across marshlands in Wales before and they had all turned out to be Buzzards. But this one had pale blue and buffish patches on wings and body - a male Marsh Harrier! Furthermore, it was flying in harrier fashion, quartering the deep vegetation a few feet above the ground, looking for the slightest movement of rodent prey. Its movement was light, more like a Kite than a Buzzard. Sometimes it rocked from side to side, countering the buffeting effect of the wind, or glided on slightly raised wings, hovering again to examine the next patch of reeds or heather. Once it landed on a small wisp of willow carr, lifting off once more at the next gust of wind. Then it disappeared, only to appear on another part of the heath a few minutes later.

A few weeks later, on the same marsh, there was a female Marsh Harrier. She looked even more distinguished than the male with her uniform chocolate brown plumage and buffish yellow head glinting brightly in the afternoon sunshine. What a pity she had not been here to meet the male bird earlier in the season! He had already apparently gone and this female seemed about to follow suit, as she now started to spiral upwards and drift slowly towards the south. There was a moment's excitement as another similar sized bird of prey descended from the skies towards the east. Could this be the missing male Marsh Harrier?

Lapwings

88

I scanned the wings through my binoculars, hoping to see the tell-tale light patches. To my disappointment, the approaching bird looked brown in a variety of mottled shades, lighter than the female Marsh Harrier but without the cream or pale blue of the male.

The new arrival of course was a common Buzzard. Harrier and Buzzard circled round menacingly, like two wartime fighter planes. The Buzzard dived headlong, the Harrier jerked rapidly to one side, talons extended. Each bird in turn plunged aggressively towards its adversary (I have never seen such behaviour between Buzzard and Kite!). Eventually the birds broke off this hostile engagement and went their separate ways.

After this anecdote we will return to the broader question of marshes in Wales. Although there are few extensive areas suitable, say, for birds like the Marsh Harrier, there are many other patches of wetland bog and reedy fields which harbour important stocks of waders such as Snipe and Curlew. The area around Trawsfynydd in Merionnydd is a good example where these two species breed along with Redshank, Oystercatchers and Lapwing. Like inflationary currency though, statistics quickly become obsolete and inaccurate, since the numbers of breeding waders are slumping into serious decline.

Near my home in Mid Wales breeding Curlew and Lapwing are hard to find as they disappear from fields they occupied only the year before. During the 1980s, a drive along the mountain road between Cwmystwyth and Rhayader would produce Curlew, Lapwing, Redshank, Snipe and Common Sandpiper, all nesting in patches of reeds or marshy ground adjacent to the road and river. All except the Common Sandpiper and maybe a pair or two of Snipe and Curlew have disappeared. The reasons for this are by no means clear since at a superficial glance most of the habitat apparently remains more or less intact. We suspect that drainage and a succession of drier summers may have had an adverse effect, together with, perhaps, an increase in sheep which gain easier access to the ground as it becomes drier. A growth in predator numbers must be significant and foxes and Crows are plentiful on the uplands. When a wader becomes scarce it will withstand little predation from species which compare in numbers or exceed it as is the case with the Carrion Crow, even though these predators must of course obtain most of their food from other sources.

Waders such as Redshank, Snipe and Lapwing prefer wet meadows, with pools and patches of reed, or long grass for nesting. A certain amount of light grazing by cattle, which leaves tufts of grass and other vegetation, improves rather than damages the environment, since none of these waders like rank overgrown marsh either for nesting or feeding. With careful management even sheep can coexist with these popular waders, providing stocks are not too high, or if only moderate numbers of animals graze the fields during the critical nesting period March to May. By then, the grass has grown long, providing rich feeding for the sheep, so there should, it would seem, be little or no loss to the farmer. Of course it is important that the meadows are left intact, undrained and free of pollutants and as far as possible kept free of predators. Perhaps a score or two of suitable meadows could be ear-marked for such management? Any financial deficit arising from conservation policies should be made good, particularly in the light of the ongoing crises in the farming industry. The welfare of rural communities obviously depends heavily upon the prosperity of the industry. The interests of wild life and farming though, can surely be managed to the benefit of both. Increasingly, co-operation between farmers and conservation organisations is taking place but government policies (and the commitment of funds) which promote wildlife and environmental regeneration in the mid and long term are vital.

The highest concentration of breeding Redshank and Lapwing in Wales is now on the Dyfi. Overall in Wales, there are few notable successes for either species. The once ubiquitous Lapwing is very thin on the ground, and almost extinguished on farmland. The British Trust for Ornithology research, as reported in *Welsh Birds* (the annual report of the Welsh Ornithological Society) for the year 2000, indicates a loss of 77 percent of the Welsh population of breeding Lapwing in only ten years. The drop in numbers is particularly catastrophic in grassland habitats. Current conservation efforts are focused on those sites holding at least ten pairs. A few hectares of suitable wetland can support healthy populations of waders when managed properly although it appears that a colony of at least this number of Lapwings is necessary to be viable in warding off the threats posed by avian and ground predators. Like other waders the chicks can run almost as soon as they are born but they need deep cover nearby for concealment. The Redshank is now rare as a breeding bird almost everywhere. Apart from the Dyfi stronghold, one or two locations on Anglesey and a few pockets along the north west coast and parts of south Carmarthenshire, account for most of the breeding sites of that species in Wales.

LAKES, POOLS AND MARSHLAND - species to look for (spring and summer):

LAKES

Great Crested Grebe	Gadwall	Moorhen
Little Grebe	Wigeon	Coot
Heron	Teal	Water Rail
Little Egret	Ruddy Duck	Black-headed Gull
Bittern	Shoveler	Black Tern
Mute Swan	Goosander	Osprey
Grey-lag Goose	Red-breasted Merganser	
Canada Goose	Pochard	
Mallard	Tufted Duck	

MARSH

Marsh Harrier	Snipe	Skylark
Water Rail	Ruff	Meadow Pipit
Lapwing	Green Sandpiper	Whinchat
Curlew	Common Sandpiper	Grasshopper Warbler
Whimbrel	Dunlin	Reed Warbler
Redshank	Little Stint	Sedge Warbler
Spotted Redshank	Curlew Sandpiper	Cetti's Warbler
Greenshank	Black-tailed Godwit	Reed Bunting

Heron

Notes: (1) Often, of course, lakes and marshland occur together, when naturally a combination of species listed above can be expected. In spring and late summer a variety of passage migrants may turn up. Time of year is important - see end of Chapter Seven for list of passage migrants.

(2) The commonest species are likely to be Coots, Mallard and maybe other species of duck such as Teal or Tufted, together with Canada Goose, Black-headed Gull, Curlew, Sedge Warbler, Reed Bunting, Meadow Pipit and possibly Skylark, depending upon the location and precise nature of the habitat.

(3) Some of our rarest birds may occur in these habitats such as Osprey, Bittern, Spoonbill Bearded Tit, and even vagrants like the Night Heron, Penduline Tit or Aquatic Warbler.

CHAPTER SEVEN
WALES IN WINTER

During late summer and autumn many millions of birds fly from northern latitudes to spend the winter in milder climates far to the south. Some pass overland or along our shores from late summer onwards, only to spend brief periods replenishing their energy with rest and food before flying off to Africa or Mediterranean regions. These are joined in their southerly flight by our own summer visitors, the vulnerable insect feeding warblers, Swallows, Cuckoo, flycatchers, Whinchat, Redstart and several others.

We have no separate chapter for migrant visitors which pass through Wales on their way to other destinations. They are dealt with under various chapters including this one. Some may stay the winter in Wales. In springtime the migration routes are northwards. Individual birds may find themselves stranded far to the west or north of their usual haunts. At the end of this chapter we provide a list of birds which regularly pass through, with a note of those months when the majority of sightings are made.

Many birders regard September and October as two of the most exciting months. This is the time when rarities turn up on coastal promontories or on offshore islands, birds which are out of the question at other times of the year. The eastern side of Britain obviously attracts more off-course rarities since it is closer to the prevailing migration routes from Scandinavia and Russia. Yellow-browed Warblers, Lapland Buntings or Red-breasted Flycatchers are much more likely to occur in Norfolk than here in Wales although such rarities do occasionally put in an appearance during autumn. The Pembrokeshire islands, Bardsey, South Stack and the Point of Ayr on the extremities of Wales, are all good places to see rare visitors. Strumble Head and Point Lynas are particularly favourable for sea watching.

Numbers of strong flying waders on a journey from Iceland or Norway pass along our western shores. Greenshank, Common Sandpiper, Whimbrel and Green Sandpipers pay fleeting visits on their journeys, while a small proportion of them may join other waders like Knot, Curlew and Dunlin and spend the winter in Wales. Look for them on tidal rivers, and the margins of inland lakes and pools.

When observed at fairly close range these waders can be separated with not too much difficulty. The three 'shanks' all have white rumps which show well in flight. Unlike the other two, the Redshank also has a characteristic bar on the rear edge of the wing. The Greenshank and the scarce Spotted Redshank are rather larger and slimmer birds, and longer in bill and leg than the Redshank. The elegant Greenshank, which breeds in the northern counties of Scotland, has green legs (and beak) as its name indicates. The sandpipers are considerably smaller. The Green Sandpiper also has a white rump while the Common Sandpiper, Ringed Plover, Dunlin and many other smaller waders, have a characteristic long and narrow wing bar. In winter the Dunlin becomes paler in plumage and loses the dark belly patch so characteristic in summer. The rather long 'drooped' bill then becomes its most diagnostic feature. If you are very lucky, you may come across a rare wader like a Grey Phalarope or even a Terek Sandpiper.

Some observers specialise in sea watching. Risking arthritic limbs and stiff joints, they perch with powerful telescopes looking far out to sea. Timing is important when watching movements of passage birds between July and November. You will easily find migrating terns off the west Wales coast as they come into shallow bays to dive for sand eels and other small fish, mostly during late summer. Sandwich, Common and Arctic Terns are the most likely. Skuas are sought after keenly by those who visit their migration routes in late summer and autumn. Their feeding habits are unique among British birds in that they pursue terns and other seabirds remorselessly until they drop the food they have caught, which is then snatched up by the pursuing bird. Watch for a raiding skua causing mayhem among the flocks of terns. Arctic, Pomarine and Long-tailed Skuas are sleek and agile birds, all having dark caps and white flashes at the elbow of the wing. The larger and more sturdily built Great Skua is about the size of a Herring Gull but is a very formidable bird. It has the 'flashes' on the wings, but lacks the long centre tail feathers of the other three species. Stormy weather with onshore gales may bring sea birds to land at different times of the year. Shearwaters, Storm and Leach's Petrels or Little Auk are some of the more exciting ones. The petrels may be recorded any time between early summer and late autumn but the diminutive Little Auk arrives from the Arctic only in wintertime.

Red-throated Divers

Unusual gulls may be encountered on passage and some may stay the winter here. Little Gulls, Mediterranean, Glaucous and Iceland Gulls are the most frequent among the rarer species. Most gulls have black primaries or wing tips. If your bird has white ones it is very likely to be one of these four mentioned species. The Little Gull is no bigger than a Mistle Thrush however, while the Glaucous Gull rivals the Great Black-back for both size and dominance. Gulls have scavenging instincts and are likely to be found in harbours or other places where offal is available.

During autumn and winter you will see other interesting birds out to sea although identifying some of them will not be easy. Winter plumage is a great levelling factor. Many waders lose their rich summer colours and much the same applies to grebes and divers. In winter they are all reduced to wintry silver and grey tones. The Red-throated Diver winters in hundreds off Borth at the mouth of the Dyfi estuary, so does the smaller Great Crested Grebe. In the south, thousands of both species are to be found in Carmarthen Bay. Sometimes there are rarer Great Northern and Black-throated Divers, or Red-necked or Slavonian Grebes. The Black-necked Grebe is much scarcer and is just as likely to be discovered on inland lakes as it is on salt water.

A telescope is useful for sea watching, its 20x or 30x magnification enabling birds to be picked out and identified over a much greater distance. As though distance and bland plumage were not handicaps enough, the observer more often than not has to contend with nature's raw elements too. Birds bob up and down in the turbulent waters, appearing tantalisingly for a second or two and then disappearing from view before a clinching identification can be made. Diving species may re-appear far from where they submerged. Poor light or misty conditions hamper visibility and cold blue hands may find it difficult to hold telescope or binoculars steady in a blustery February wind.

To many birders, winter means wildfowl and waders. Millions of ducks, swans and geese seek refuge on Britain's lakes, mudflats and estuaries during the coldest months. Most of them descend wearily on our waters after travelling thousands of miles from Greenland, Iceland, Scandinavia and Russia. It is said that the Whooper Swan travels from Iceland to Scotland non-stop in fourteen hours, travelling at 27,000 feet at a temperature of $-42°$c. This amazing feat, if true, is probably paralleled by other wildfowl. The Whooper Swan and smaller Bewick's Swans are easily distinguished from the common Mute Swan by their black and yellow bills, and when floating on the water their necks are usually held more upright. Both species are seen in small numbers in various counties of Wales.

Geese are few in number compared with the huge skeins which can be seen flying against a winter sunset on the Wash or on the Ribble estuary in Lancashire. There are however a few sizeable flocks of geese regularly wintering in Wales. 150 Greenland White Fronts annually return to the same fields in the same part of the Dyfi estuary as they have done for many years. The Russian race of the species is traditional at Dryslwyn in Carmarthenshire but has virtually disappeared. Sometimes there are a few Barnacle Geese beside the river at RSPB Ynys Hir and small numbers of Brent Geese feeding on the salt flats of the Dyfi Estuary. Grey-lag geese, many descended from locally bred feral birds and identified by their orange bills and legs and light grey scapula patches, can be seen over various parts of Wales. The Pink-footed Goose occurs most often in North Wales at places like the Glaslyn Marshes near Porthmadog or on the Dee estuary but this bird, so common on the Ribble, is surprisingly scarce in Wales. Note the small beak with pink tip and darker

Goldeneye

neck of this species compared with the Grey-lag Goose. The Brent Goose winters in reasonably good numbers in Wales, its stronghold being around the Burry Inlet and this is probably the most common of the truly wild geese. The most familiar goose though and perhaps the least exciting one, is the Canada Goose whose increasing feral populations are to be seen at all times of the year.

There are some interesting duck which you may see out on the sea. The Common Scoter in its dark plumage (male is black) is the commonest and easiest to identify as rafts of black dots appear and disappear in the swell. The white wing bars of the similar Velvet Scoter can be seen when the bird is bobbing up and down on the waves but are more easily noticed when this scoter is in flight. The male Scaup is similar to the Tufted Duck but has a light greyish back, while the females are distinguished by a distinctive white shield above the upper mandible of the Scaup. There is usually a scattering of Eider at various places on the Welsh coast but the two main sites are Burry Inlet and Dysynni in Meirionydd. Perhaps the most exciting of our wintering duck is the Long-tailed Duck. The black and white male with his long tail and dark cheek patch is a lovely bird but you will be lucky to encounter this species. It is most likely to be seen in shallow estuarine localities. Its only rival in the glamour stakes is the Smew, another species from the Arctic (the male is also predominantly white) which more often winters on the lakes of Glamorgan than elsewhere. It is a regular visitor to Eglwys Nunydd reservoir next to the Port Talbot steelworks.

There are a number of particularly well favoured areas for seeing large numbers of duck in winter where perhaps eight or ten species may be seen. A good place is the Wildfowl and Wetlands Trust at Penclacwydd on the Burry Inlet where you have the added advantage of being able to observe the birds at close quarters from the comfort and camouflage of well placed hides. Despite terrible weather I was able to see nine species of duck, Shelduck, Mallard, Teal, Gadwall, Pintail, Wigeon, Shoveler, Pochard and Tufted Duck on a very bleak December afternoon. There were also a number of Whooper Swans and several hundred Brent Geese flying over the estuary, while scores of Lapwing and a flock of 200 Black-tailed Godwits probed in the mud. I was told by a group of birdwatchers that a rare vagrant from America, the Long-billed Dowitcher, had been seen among the Godwits. Unfortunately, if it was still there, we were unable to detect it. Three Little Egrets were conspicuous in their snowy white plumage as they probed about in the brackish mud, a small contingent of the 50 or so egrets which now regularly winter in Wales.

The Burry Inlet, in close range of both Swansea and Llanelli, has extensive salt flats and pools and together with Carmarthen Bay to the west is arguably the best area of all for wintering wildfowl in Wales. The Cleddau in Pembrokeshire attracts large numbers of duck and waders, while the mouth of the Ogmore, near Bridgend, is noted for Goldeneye. In Mid Wales the richest areas for bird life at this time of the year are the Teifi and Dyfi estuaries and their hinterland marshes, while north of the Dyfi, the Mawddach estuary and Aberdysynni (near Tywyn) are notable sites though on a modest scale. Further north Glaslyn is important, together with Dinas Dinlle, Traeth Lafan near Bangor and the estuaries of the Dee and Conwy. In recent years unfortunately, industrial development has marred and polluted the Dee and made public

Pintail (left) and
Wigeon (both males)

access more difficult. This outstanding area though, still attracts large numbers of Knot, Black-tailed Godwit, Grey Plovers and other, commoner wintering waders such as Dunlin and Oyster-catchers. There are good sites to be recommended

o n Anglesey, the so-called 'Inland Sea' near Holyhead, Llyn Alaw the largest lake on the island, and the coastal marsh at Malltraeth, being among the best.

One of the commonest duck on our estuaries in winter is the Wigeon, whose presence is often signalled by its distinctive whistling call which echoes through the fog or mist even when the birds themselves are invisible. In flight, the broad white wing patch of the males is a prominent feature which enables a flock of Wigeon to be identified at some distance. The fast flying Teal, a small duck with narrow wing bars noticeable in flight, can appear almost wader-like as a group of them twists and turns with great agility over the marsh. This species is usually found in good numbers, especially where there are sheltered pools. Here small flocks of Teal can be quite inconspicuous as they huddle together close to the reeds. The long neck of the Pintail, and particularly the distinctive head pattern of the drake, will be easier to pick out than his pointed tail when seen at a distance. Wigeon, Teal and Pintail, like the Mallard which is also common on brackish estuarine pools in winter, are all surface feeding ducks.

The Red-breasted Merganser is another familiar bird on the estuaries at this time of the year. With its long slim neck and fine bill, this species reminds one more of a grebe or a diver than a duck as it slips beneath the water, hunting fish in the deeper parts of the river. Pochard and Tufted Duck are usually observed on fresh water lakes either near the coast or inland. Some upland waters which are wildlife deserts in summer often hold small numbers of these duck in winter but the largest concentrations of them are to be found on low-lying larger lakes and reservoirs. These two species together with the Goldeneye and the 'sawbills' all dive for their food, and this habit provides a good clue when factors of poor light and distance make identification difficult.

The male Pochard, with his soft grey back and red head, is easy enough to identify. So is the male Tufted Duck, provided care is taken to separate him from the male Goldeneye which has more white showing, a distinctive white

Grey Wagtail (male)

Little Ringed Plover

Peregrine

Redshank

Great Crested Grebe

Snow Bunting (female – winter)

Red Kite

Buzzard

spot below the eye, and broad white wing bars noticeable in flight. I know of sizeable upland lakes close to my home where the only birds to be expected in winter are two or three Goldeneye. Although found on lakes this species has no objection to salt water and occurs in reasonable numbers on some estuaries. The female, in fact, is actually the one with the golden eye and her brown head and ring make her more distinguished and easier to identify than most of the other rather featureless female ducks. Their sombre colours provide vital camouflage when they are incubating eggs. Such precautions are not necessary for those species like the Shelduck, Goldeneye or Goosander which lay their eggs in the darkened tunnels of banks, or holes in trees.

While spring or autumn is the time for the passing rarity, winter is the time for the spectacular show. Vast squadrons of ducks or geese fly in 'vee' formation against a bronze December sky while huge flights of Dunlin or Knot twist and turn over the estuary in perfect unison, with all the skill and harmony of an air display team. These and other waders descend in force on watery places free of the fatal frosts they would endure if they had stayed in their summer quarters. Generally speaking the largest numbers are to be found in similar locations to the wildfowl, given the proviso that, as a rule, they require more mud and less water! Estuaries are therefore particularly important for waders. Dunlin gather in their thousands on the Dee, the Cleddau estuary and Burry Inlet, where there are also massive invasions of Knot, Lapwing, Oyster Catchers and Curlew. Smaller numbers of Ringed Plovers, Black or Bar-tailed Godwits and Redshank can also be seen on saltmarsh or beach, while the dainty, silvery Sanderling runs up and down with the advancing and receding tide looking for freshly exposed sandworms. Short-billed, medium sized waders are likely to be plovers; Lapwing, Golden or Grey Plover. Look for the gold-flecked plumage of the Golden Plover, which is sometimes to be seen in large flocks on beaches or on adjacent fields where they can be surprisingly inconspicuous. Both this species and the similar Grey Plover look pale reflections of the handsome birds they are when seen to their best advantage in summer. In flight the Grey Plover shows a distinctive white rump and white wing bars which the

'Shoreline waders'–Knot (left) Grey Plover (centre) Bar-tailed Godwit (right)

Golden Plover lacks. Both species fly fast on pointed wings so different from the broad ended wings of the Lapwing.

The Snipe and the rare Jack Snipe feed quietly among the short reeds, probing in the mud for the invertebrates upon which their survival depends. When the ground is frozen over the Snipe is in trouble, and may have to seek patches of frost free mud near rivers and brooks well away from its usual feeding areas. If startled it flies steeply from the ground, twisting and turning until it is a safe distance away. The Jack Snipe is reluctant to fly, and when it does its path is straighter than that of the erratic Snipe and it usually loses no time in landing again. Its habit of 'freezing' means it is often overlooked. The Woodcock is like a huge plump Snipe which is normally found in wooded locations. In winter it is much more common in Wales than in summer and birds may be flushed in fairly open locations beside pathways or in damp field-side copses. In hard weather when the land is frozen solid, Woodcock, like Snipe, must make do with whatever soft ground they can find for feeding. Any large brown bird which flies up suddenly in front of you from the woodland floor or even a roadside ditch, and disappears silently and rapidly from view within a matter of seconds, is likely to be a Woodcock.

Foraging among stones and rocks on the beach you may find one or two other species joining the conspicuous Oystercatcher in searching for a meal. Watch the common and aptly named Turnstone flipping pebbles and stones to expose crustacea beneath. This is a darkish looking small wader which shows an intricate brownish-black and white pattern in flight. On some rocky shores you will come across the Purple Sandpiper which as its name implies, has purplish brown plumage. It is pale underneath and has a yellowish bill and legs. I have seen them in small groups probing among the rocks on many occasions at Aberystwyth where they overwinter each year and they occur in small numbers at other places around the Welsh coast. They blend well with the black rocks and are most easily seen when the tide just begins to recede, exposing only those rocks close inshore.

Before we leave the coast and head inland, we should mention that the wealth of bird life on estuaries during the winter acts as a magnet to birds of prey. At Cors Fochno in north Ceredigion for example, Merlin, Peregrine, Hen Harrier and more rarely Short-eared Owls, can all be seen over the marshes in winter. On one occasion I observed an Owl and Peregrine circling round, climbing higher and higher into a grey bronze late afternoon sky, seemingly in some form of mock combat.

Throughout the seasons, except when breeding, there is a constant movement of birds between one country and another and within these islands, since there is a perpetual struggle for survival against cold and starvation. Although Wales is too westerly to catch the bulk of passage migration between north and south, it does have the advantage of a mild maritime climate. During very cold spells birds may be driven westwards, starved of nutrition by deep snow, frozen lakes and impenetrable icy ground. The Bittern sometimes finds shelter in deep reedbeds where it is a star attraction with birdwatchers who will try to catch a glimpse of this retiring bird. The Bearded Tit may also winter in western reedbeds, and in 1996-1997 up to three rare Penduline Tits spent the whole season at Kenfig Pool in Glamorgan. This is a species which does not breed in

Britain but it is rapidly expanding westwards in Europe and could do so within the next decade. It weaves an amazing inverted, bottle shaped nest which hangs from the slender stems of riparian vegetation.

A survival strategy of many passerines such as larks, finches, thrushes and tits is to move about in flocks outside the breeding season, whilst others, like Rooks and Starlings, roost together in huge communities. If you stand on the promenade at Aberystwyth during late afternoon in winter, squadrons of Starlings in groups of up to 500 or more will pass by on their way to roost noisily on the girders of the pier. Black specks emerge from a darkening sky, all flying from a northerly or easterly direction as though guided by air traffic control and disappear into the gathering dusk. The passage takes probably 20 minutes and more before the last bird is settled on the crowded rafters, and the total congregation numbers several thousand. Where do all these birds come from? Recently I have seen Starlings clinging like huge insects to the roof tiles of a large church like some horror from an Alfred Hitchcock movie. Even the Carrion Crow, a solitary species in summer, will roost in trees as Jackdaws do. Close to my home as many as 50 or 60 Ravens gather towards dark to roost on electricity pylons and tall spruce trees.

Flocks of thrush-like birds feeding in the fields or greedily stripping berries from hedgerow trees between October and early April are almost certain to be Fieldfares or Redwings. Often the two species occur in mixed flocks and are our most familiar winter visitors. The larger Fieldfare (the size of a Mistle Thrush) is the more conspicuous as he utters loud 'chack-chack-chack' sounds and reveals a colourful plumage of rich red-/brown back and wings, blue/grey head, rump and tail, and buff underparts. Look closely at the Redwings in the flock. They are neat attractive birds, a little smaller than Song Thrushes, with white eye stripe and rouge flanks just below the wing. Our two resident thrushes and the Blackbird are also augmented by inward immigration from the continent but they are rarely to be seen in large flocks.

Some species, although present at all times of the year, are more easily observed in winter than at other times. Resident Crossbills, Siskins, Bullfinches and Starlings are supplemented by larger numbers of visitors from elsewhere. Harsh conditions, and the urgent need for food, bring Coal Tits and Goldcrests down from the canopy of evergreen trees. Delightful Long-tailed Tits trill from bare branched bushes and Redpolls hang upside down from the crowns of birch and alder, prizing out the small seeds from their pods. At this time of the year, when the branches of deciduous trees are bare, birds are more easily seen. The Bullfinch, one of our most attractive small birds, is usually heard before he is spotted among brambles, bushes and thorny trees. If the male

Great Grey Shrike

99

emerges into view, his contrasting pinkish red breast, black cap, grey back and blue/grey wings and tail are striking. The female is similar but her colours are more subdued. The Bullfinch is one of the few small birds which has a very distinctive white rump. The other two which do, House Martin and Wheatear, do not arrive until spring.

If there were more birdwatchers in Wales there would undoubtedly be more records of scarcer species. In November I drove onto a car park near the sea at Borth and was amazed to see two Shore Larks busily foraging for food in the grass and shingle. Modestly unaware of their stardom they continued to potter about unobtrusively, too intent on their search for grubs and seeds to bother about me as I watched in fascination from the car. Fortunately I had binoculars handy and was able to enjoy close up views of the attractive black and primrose-yellow head pattern of this unusual species. Black Redstarts occur regularly in winter at various coastal and urban locations and in places such as old building sites, ruins and waste ground. This species is alone among small birds in being predominantly black/dark grey, but it has the same orange tail flash as our more familiar Redstart which is only seen in summer. In winter the plumage of both sexes is smoky-grey. Black Redstarts have bred in Wales on a number of occasions. They are urban nesters, preferring derelict sites and pairs are most likely to be discovered in towns and cities in South Wales. Close inspection of flocks of 'Linnets' may occasionally prove to be Twites if they have plainer colours with pinkish rumps.

A short while ago I stood talking to two birdwatchers who told me they had been fortunate enough to observe a Great Grey Shrike at Cors Caron in October. When discovered sitting prominently upright on a bush or post, this immaculate grey, white and black bird, the size of a thrush, will attract a lot of enthusiastic attention. It is most likely to be found on heather covered scrubland or bogs with a good scattering of bushes. They also confided that three Firecrests had been seen near the public path. These finds show what can be discovered by dedicated enthusiasts who visit selected sites on a regular basis. Firecrests breed locally in various woods in southern Britain, notably the New Forest and parts of the home counties, and a few pairs used to be found in the Wentwood Forest and by Lake Vyrnwy, but unfortunately, as reported in Chapter Four, they are no longer to be found there. As we suggested there, look at the 'Goldcrests' carefully if you are able to get a good view. If your bird has a bold, black-bordered white eyestripe above and below the eye, you are looking at a Firecrest!

Whereas Firecrests are usually passage birds, most often discovered in autumn or spring, Bramblings normally stay throughout the winter. They appear rather like Chaffinches at a casual glance, and may occur singly or, more often, in small flocks. Bramblings have a liking for beech mast so if you see a flock of 'Chaffinches' feeding beneath a stand of beeches, scrutinise them closely. The Brambling is an orange colour on the breast where the Chaffinch is dull pinkish. The shoulder patch of orange-brown is particularly noticeable, together with the dark grey head (male) and bolder black and white lines on the back.

Bramblings are not rare, but there are other small winter visitors which cause a stir of excitement whenever they show up. The Snow Bunting is easily identified by its brown-flecked, whitish plumage. Like the rarer Shore Lark, it is most likely to be seen near the coast but sometimes it is found in the uplands. Only a few stoic people look for birds on the hills in the colder months, so the Snow Bunting may occur more often in Wales than the dozen or so annual records suggest. Recently a friend told me of one which flew in front of her car on the high road approaching the moors near Penrhyncoch, Aberystwyth. A flock of these buntings are like drifting snowflakes, and quite unmistakeable. Unless you are lucky enough to see the male in summer plumage the Lapland Bunting is not so easy to identify, since otherwise it must be distinguished carefully from the female Reed Bunting. This species is also worth keeping an eye open for during a walk in the vicinity of dunes, rough pasture, or headlands. For instance, they have been seen on more than one occasion in recent years at Great Orme's Head. Other rare passerines which could be overlooked are possible during the autumn, birds like the Richard's Pipit, the

Brent Geese

Yellow-browed Warbler and the Bluethroat, which has a liking for marshy sites.

If I had to pick a gold star bird from our rarer winter visitors I'd probably choose the Waxwing. It is far from common in Wales, occurring much more often on North Sea coasts but in 1997 there was an influx. No less than 128 birds were reported by lucky observers between December and March. The Waxwing usually occurs in flocks and in flight has a silhouette rather like a Starling; but there the similarity ends. The Waxwing has an unmistakable pinkish chestnut crest, black eye-stripe and throat patch. The plumage is generally chestnut above, with grey rump and pinkish brown underparts. Most exotic of all, it has scarlet waxy markings on dark wings which at their tips are 'pencilled' with white and yellow. As an extra touch of extravagance, the tip of the tail is a vivid yellow. Waxwings are usually seen feeding on berries in tall hedgerows and gardens in the depths of winter.

About the end of March or April the return migration begins and there is an exodus of all the wildfowl, waders and other birds which have stayed with us over the winter months. From then until well into May, there is a northward movement of birds from the warmer south. Many of them will be the same birds that travelled the same route six months previously but in the opposite direction. At this time, Ospreys may be seen plunging for fish in Welsh lakes and rivers which played host to them the previous August or September. Some species will follow a different route on the return migration. Whimbrel for instance are much commoner in spring than they are in autumn. They are smaller than Curlews, with shorter, stouter curved bills and a noticeable eye stripe. The top of the head is soft warm brown rather than striated as it is in the case of the Curlew. The two species can be confused if you are unfamiliar with them. I remember a friend remarking 'Look at that Chough standing by those Curlews'. It was a Chough all right, but the 'Curlews' were Whimbrels!

As the wintering birds fly to the north, our eagerly awaited new arrivals return to herald the coming summer. Wheatears are recorded at coastal sites early in March and Chiffchaffs are heard singing well before the end of the month. By mid April, the first Blackcaps and Willow Warblers have arrived to swell the chorus of bird song provided by resident Wrens, Blackbirds and Robins. Later in the month the early migrants are joined by Redstarts, Pied Flycatchers and Wood Warblers. Whether birders or not, we watch for the first Swallow skimming over the farmyard barn, or listen joyfully to the Cuckoo calling from a distant meadow. By May the countryside is full of bird song and many species are making final adjustments to their nests or incubating clutches of eggs.

WALES IN WINTER - species to look for:

ON THE SEA
Divers (3 species)
Grebes(4 species)
Storm Petrel
Leach's Petrel
Cormorant
Shag
Auks
Gulls
Scoters (2 species)
Scaup
Goldeneye
Long-tailed Duck

ON THE SHORE

Oystercatcher
Lapwing
Golden Plover
Grey Plover
Ringed Plover
Turnstone
Purple Sandpiper
Dunlin
Sanderling
Curlew
Bar-tailed Godwit
Gulls
Rock Pipit

ON MARSH & ESTUARY
Heron
Little Egret
Bittern
Geese (6 Species)
Wigeon
Teal
Mallard
Pintail
Shelduck
Birds of Prey
Short-eared Owl
Black-tailed Godwit
Bar-tailed Godwit
Spotted Redshank
Redshank
Greenshank
Curlew
Ruff
Snipe
Jack Snipe
Green Sandpiper
Knot
Dunlin
Oystercatcher
Ringed Plover
Golden Plover
Lapwing
Water Rail
Meadow Pipit
Skylark
Stonechat

ON LAKES
Whooper Swan
Bewick's Swan
Mute Swan
Canada Goose
Grey-lag Goose
Goosander
Red-breasted Merganser
Smew
Goldeneye
Pochard
Tufted Duck
Most other duck species
Coot
Moorhen
Great Crested Grebe
Little Grebe
Great Northern Diver

ON LAND

Woodcock
Redwing
Fieldfare
Firecrest
Great Grey Shrike
Black Redstart
Twite
Brambling
Snow Bunting
Lapland Bunting
Shore Lark
All resident land species

Notes: (1) The commonest species at sea are gulls and Cormorants while on the shore Oystercatchers, Dunlin, Turnstones and Ringed Plovers are the most frequently seen. Locally, Knot, Grey plover and godwits may occur in sizeable numbers.

(2) In estuarine habitats Oystercatcher, Curlews, Redshank and wildfowl such as Wigeon and Teal may be plentiful. Birds such as Lapwing, Curlew and wintering geese are often found in numbers on coastal fields, beaches and marshes as well as estuaries.

(3) Pochard, Tufted duck, Teal, Wigeon and Goldeneye, together with Coots, Mallard and Canada Geese are among the most familiar wildfowl to be seen on lakes in winter.

(4) The above groupings of birds according to habitat is a guide only, since a number of species may occur in several habitats. The Red-breasted Merganser for example may be found on the estuaries, on lakes, or on the sea. Swans, geese, Wigeon, Teal, and other species may be seen on all kinds of watery and marshy habitats.

(5) The above lists are not comprehensive, and a much larger range of birds may be seen in various habitats in winter including, of course, all of our resident birds.

PASSAGE MIGRANTS THROUGH WALES

Marsh Harrier	—	April May - July Sept., some in winter
Osprey	—	April-Oct. - all months especially Aug. & Sept.
Spoonbill	—	April May June - some winter
Mediterranean Shearwater	—	Aug.- Nov.
Sooty Shearwater	—	July-Oct.
Leach's Petrel	—	Sept. and Oct.
Pomarine Skua	—	May and Aug-Oct.
Long-tailed Skua	—	Aug until Oct.
Arctic Skua	—	July-Oct.
Great Skua	—	July-Oct., some in winter
Mediterranean Gull	—	All months
Little Gull	—	May and Aug-Oct also winter
Black Tern	—	Aug-Oct. also spring
Sea Terns	—	Aug-Oct. also spring
Garganey	—	April-May and Aug-Sept.
Avocet	—	Mainly April and May
Dotterel	—	April and May, some in Aug. and Sept.
Little Stint	—	Mainly Aug. and Sept., some Oct.
Curlew Sandpiper	—	Aug., Sept. and Oct.
Ruff	—	All months, especially April, Sept. and winter
Black-tailed Godwit	—	Throughout the year
Whimbrel	—	April until Aug., mainly spring
Spotted Redshank	—	July until Oct. but all months
Greenshank	—	July-Oct., but all months except June
Green Sandpiper	—	Mainly late summer, some winter
Wood Sandpiper	—	May and late summer. July-Sept.
Hoopoe	—	Mainly spring, occasionally late to Oct. and Nov.
Woodlark	—	Chiefly Oct., a few winter
Richard's Pipit	—	Rare, usually Oct.
Black Redstart	—	March-May and Oct-Nov., some winter
Turtle Dove	—	May - breeds Gwent only
Firecrest	—	Oct-Nov., also spring, a few winter
Golden Oriole	—	May and June

Note: Some wintering species such as Sanderling and Dunlin are at least as well known as passage migrants.

CHAPTER EIGHT
BIRDS OF PREY IN WALES

We have already touched upon most species of raptor in the various sections of the book, but birds of prey deserve a section to themselves. To begin with, they wander in search of prey so may occur in a variety of habitats. The Peregrine for example, whose very name means wanderer, will nest on coastal cliffs or inland crags; it will hunt over farmland and forest, and in winter will dive among frantic flocks of waders on river estuaries. Another reason for highlighting this group of birds is because they are so strongly represented in Wales.

I remember a few years ago, driving round Mid Wales on the last day of March with two companions, seeing how many birds of prey we could count in one day. In valleys where there was no through road, we counted only birds seen in one direction. Our total for the day was 126 birds, comprising 17 Kites, 5 Peregrines, 5 Kestrels, 2 Goshawks, 1 Sparrowhawk and 96 Buzzards. Where else in Britain, and even in most parts of Europe if it comes to that, could you see this many raptors? With a little luck at that time of the year we might have added Hen Harrier and Merlin to the list. Many of the Buzzards and Kites were circling hilltops, sometimes in groups of 10 or 12, rising in the warm afternoon air.

The star bird of prey, of course, is the Red Kite although the introduction of the species into various parts of Scotland and England, where there are now 80 breeding pairs, has rather taken the gloss off its celebrity status in Wales. Here the Kite has gone from strength to strength in the past 20 years. When we wrote *In Search of Birds in Mid Wales* in 1988 there were 40 pairs. In 1998 there were nearly five times that number. The range has expanded mainly south, although there are breeding records from Caernarfonshire, and in 1997 two pairs of 'Welsh Kites' were reported to have bred in the English border counties. In winter, Kites are fed regularly at several places and between October and March you may see up to 50 or 60 birds together. The Kites rarely land to pick up pieces of meat. Instead, using great skill, they dive and dispossess the more ponderous Crows barely inches from the ground. Presumably this method of feeding is energy efficient.

Red Kite

Watch the Kite carefully in flight. The wings and tail are longer than those of the Buzzard and the plumage is a striking combination of tones. The body is generally ruddy brown, the head whitish, and there are patches of black and white on the wings, while the tail has an orange-brown translucency in sunlight. Observe the Kite swivel its tail like a rudder to change direction or gain balance. The forked tail, of course, is a give-away. Yet even when you can't see the forked tail or the rich colours, the Kite has something else - buoyancy. This magnificent bird flies with a consummate ease and precision that distinguishes it from the lumbering Buzzard, which only looks graceful once it is able to glide or soar on the uplifting thermals. No wonder the toy kite was named after the bird, since the air is this raptors perfect element. Neither does the Kite so often receive the nagging attention of Crows that are so often seen plaguing the lives of Buzzards which have strayed too near the corvids' nests. When it does, the Kite is usually able to outmanœuvre and shake off its tormentors.

Two questions come to mind when considering the Red Kite's precarious status in Wales during the twentieth century. Why did the Kite survive here when it became extinct in England and Scotland? It does, after all, fare better in drier, gentler, more pastoral settings than prevail in most of Wales. For this reason it has never been common in North Wales. The reasons for the Kite's survival in Mid Wales lie in the remoteness of the country combined with the relative absence of game-keeping. Then there is the sympathy of the farming community and the unstinting work of individual naturalists and conservation organisations, without which the Kite would have become extinct here long ago. The other question is 'Why has the Kite been so successful in the past two decades, when it had teetered on the brink of extinction for the previous 80 years?' Less persecution and better protection are two answers. We still hear of nests being robbed and birds killed by poisons intended for foxes and Crows, illegally impregnated into dead sheep. The Kite has been guarded at its nesting sites for a long time with the help of farmers but perhaps the conservation message has gained in strength. The Welsh Kite Trust, set up in 1996 is now continuing the work of protection inspired by the efforts of Professor J. H. Salter in the early part of the century, and later sustained over many years by the Kite Committee (an organisation headed by the RSPB and the Nature Conservancy-now the Countryside Council for Wales). As the Trust points out, despite a recent increase in numbers the Kite is still a rare bird. It has the potential to be common but to put its current position in perspective, there are 23,000 people and 100,000 sheep for every breeding pair of Kites in Wales! What an asset it would be to the countryside if this outstanding species could one day return to the abundant status it enjoyed long ago.

Winter feeding may reduce mortality rates. For many years a lady charged a small fee for the privilege of watching the Kites at a site three miles south of Tregaron, the money being donated to the restoration of a church in the town. That site has now been replaced by another at Pont Einon, just outside Tregaron on the road to Bronant. Feeding of Kites at this and other sites usually takes place during the early afternoons, October till March. The three other main winter feeding stations are currently at Gigrin Farm just south of Rhayader on the A470, Penlan House, Trefilan, Talsarn, and Nant yr Arian, the forestry site west of Ponterwyd (A44). A journey to one of these sites is well worth while since there are few more fascinating experiences than watching a large gathering of this long-winged rakish raptor in the air at the same time, reminding one of the scenes which our ancestors must have witnessed in mediaeval times.

As we approached Tregaron some while ago, the frosty-pink January sky seemed to be filled with them as they circled overhead, or flew with deep and deliberate wing strokes over the marsh. We counted forty Kites in the air and a greater number of Crows which were mostly gathered around the offal lying in a rough roadside field at the edge of the bog. Reflecting the strong sunlight in flashes of black, white and ruddy brown, the Kites swooped down into the cluster of Crows, twisting and turning with great agility. Almost touching the ground, the large raptors would often emerge from the fray with morsals of meat clutched in their talons. Not always were they successful though, and on one occasion an indignant Crow managed to retaliate, causing a startled Kite to drop its prize back into the jostling crowd of birds. The 'strikes' never involved more than two or three Kites at any one time, as though each was courteously waiting to take its turn 'at the table'. The action was so fast that in the absence of a 'match of the day' type action replay, it was difficult to see clearly how the Kites actually prised the meat away from the Crows. The small gathering of people stood coldly in their overcoats and anoraks, enthralled by the whole performance. Within half an hour the meat was all gone and the Kites had mostly dispersed, soon to be followed by their human admirers who had been treated to a memorable experience. One or two ardent birders stayed on until the dusk settled on the rusty winter marsh, vainly hoping to catch a glimpse of the single American Wigeon which had been reported browsing among the flocks of ordinary Wigeon on the banks of the river.

One interesting theory to account for the recent breeding success of the Red Kite is that the inbred population benefited from the genetic contribution of a female German Kite which wandered into Mid Wales and bred with local Kites more than two decades ago. Recent DNA analysis has shown that at one time during the 1930s the handful of remaining pairs managed to produce merely a single female offspring, indicating that the Kite escaped oblivion only by the 'skin of its beak'. When it was almost on the point of extinction in the early part of this century, the last few pairs lingered on in the district of Rhandirmwyn in the upper Tywi. There may also have been a pair or two further north and eventually the focal point for breeding Kites shifted to Ceredigion. Twenty years ago a farmer at Cwmystwyth told me that his father used to leave the carcass of a dead sheep on the field for his Kites, against all Ministry of Agriculture regulations!

Kites usually nest in large trees, not in the stunted sessile oaks which hang thickly at the sides of steep valleys. Groups of trees with ample space between them are usually preferred. The favourite tree is oak whilst another popular choice is beech. I have seen one from the road quite low down in an alder tree, where the sitting female protruded conspicuously from both sides of a surprisingly small nest. Sometimes a conifer is chosen, but less often than in the case of the Buzzard. After several years use a nest may become thick and bulky, almost like that of an Osprey. The pairs are most conspicuous over their nesting woods in March, and the two or three eggs are usually laid in April. If you inadvertently approach a nest too closely, the Kites will circle overhead. This is

Goshawk

the time to retreat as quickly a possible, since this species is still rare and must be given every chance to hatch and fledge its young successfully.

Comparisons are often made between Kites and Buzzards, probably because they are of similar size, frequently soar together over hills and valleys, and share similar habitats. Surprisingly there seem to be few disputes between the two species, especially when you consider that they have apparently competing food and nest site requirements. The Buzzard is a large, robust, powerful bird, though it lacks speed and has no reputation for manoeuvrability. I was surprised therefore to see one warding off the unwelcome attention of a Crow by rolling rapidly and repeatedly through 360 degrees, a feat of aerobatics I have never seen any bird perform before. It is an opportunistic feeder that will take a wide variety of prey. Rabbits and rodents spotted from high above the ground may be caught following a steep and rapid glide. The Buzzard may hover when scanning the ground for its quarry which its telescopic eyes can pick out at a great distance. It lacks the fine skills of the Kestrel, but usually hangs stationary against the wind near the brow of a hill. The Buzzard is not too proud to forage for earthworms in a freshly ploughed field, nor to feed at the carcass of a dead sheep, taking precedence in the pecking order over Kite, Raven or Crow. One of the delights of birdwatching in Wales is to spot a huge and tawny Buzzard perched on a telegraph pole or fence post, perusing the ground from its high perch for a flicker of movement below. Despite its limitations where pace and agility are concerned, the use of these various methods makes this impressive bird a successful hunter, and as if to underline the point it is our commonest bird of prey. The Buzzard is more numerous in Wales than any other part of Britain, where although it is still concentrated in western areas there are clear signs that it is breaking out of its strongholds and starting to recolonise lowland districts in the east.

The Buzzard usually nests in a wood but sometimes will chose a hedgerow tree in the manner of a Crow. Unlike the Kite, it will not infrequently build its large bulky structure on the branch of a larch, Douglas fir or other conifer. In many ways the Buzzard has a claim to rival the Kite as Wales's national bird. Among the omnipresent corvids and other common birds of the woods and valleys, the aquiline Buzzard cuts an imposing figure. It is ubiquitous throughout Mid and South Wales, penetrating as far even as the suburbs of Cardiff. It is scarcer in those parts of North Wales where the mountain terrain suits it less well than the soft green wooded valleys.

The third of the three 'heavyweights', so to speak, is the Goshawk. Numbers of this ferocious hawk have increased in leaps and bounds since it first escaped, or was released into the wild by falconers, back in the 1960s. It is found in parts of Scotland, the Kielder Forest in Northumberland, and other parts of England including the Welsh border counties, the New Forest and the south-west. The patchwork of woodland and open moorland so typical of Wales seems to suit the Goshawk particularly well. To begin with it was confined to the larger forests, but now occurs even in relatively smaller woodlands. The female, a much larger bird than her mate, normally lays three to five eggs (occasionally less) compared to two or three laid by the Kite or Buzzard. The large untidy nest is usually supported by a sturdy side branch close to the trunk of a mature larch tree and is often located in a sheltered part of the wood. The birds need plenty of room to enter and leave the nest without damaging their wings, so trees too closely set together are of no use.

I remember once stumbling upon a large nest in Mid Wales which I dismissed as a very old Buzzard's nest. It was easily visible below me as I walked along the forest path and looked bulky and misshapen, rather like an old mattress. I failed to give any significance to a fresh sprig of spruce which embellished this untidy mound of sticks. When I returned to the site two months later, there were three fluffy Goshawk chicks visible from the track above the nest slope.

The Goshawk is most easily observed from February to April when a pair may be seen soaring and displaying over their nesting woods. The female may look nearly as large as a Buzzard and in fact she is longer, but her 'shape' is different. She has shorter wings and a longer tail than the more moth-like Buzzard which often soars with wings held in a shallow 'V'. The Goshawk's wings beat rapidly, although the stroke is often shallow. A few wing beats are usually followed by a glide on stiff wings and sometimes the bird will dive at great speed into a wood.

The outline of the Goshawk and flight mode is similar to that other accipiter, the Sparrowhawk, but even the male Goshawk is considerably larger and thicker-set than the female of that species. A useful point is that the larger bird has a more round-ended tail than the Sparrowhawk. Both sexes have a grey-brown look and appear whitish around the thighs. In rapid flight the wings may seem pointed and from certain angles Goshawks may appear more like Peregrines than any of our other birds of prey.

The best chance of seeing a Goshawk is to take a walk along a ridge where there are extensive stands of mature conifers and good views over the valley. The time of day is not especially important. Good weather always helps but they can be seen in gusty weather and rain. Look skywards; there may be a Goshawk soaring, perhaps in the company of Ravens or Buzzards. If you are lucky, you may see a pair displaying, chasing each other over tree tops, or diving and climbing steeply above the wood.

The formidable Goshawk is capable of taking almost any size of bird up to a cock Pheasant. Its hunting prowess and macho style make it popular with many people who love birds of prey. It takes Wood Pigeons and its propensity for corvids enhances its popularity even more. Some years ago, when walking along a picturesque valley of mixed woodland and bracken-clad heath in Mid Wales, a Magpie appeared overhead as I approached a larch plantation. Suddenly a large hawk swooped upwards from below the tree canopy, clutched the Magpie in its talons, and with a flurry of feathers both birds disappeared back into the wood. The Goshawk is less exclusively a bird feeder than its smaller relative the Sparrowhawk. I watched one circling a wooded valley bordered by open moorland not far from Newbridge last April. Slowly the bird circled above me and then swept into a dive, hurtling towards the ground at high velocity before braking rapidly to take a rabbit on the steep hillside. The Goshawk frequently hunts over open ground but also is said to take many squirrels and birds from the cover of trees.

There was a time when Goshawks could only be expected in the larger forests of Wales. There are now an estimated 300 pairs, more than in the rest of Britain together, and woodlands of smaller size are now frequently used. This could be bad news for the Sparrowhawk which appears to be scarcer in those woodlands where Goshawks are to be found.

Like the Buzzard, the Goshawk is now widespread across Wales, so you may well be able to find them in your area provided there are adequate woodlands. They are least likely to be encountered at the time of writing in Anglesey, Pembrokeshire, South Glamorgan and the Lleyn Peninsula, all areas on the extremities of Wales containing little extensive woodland.

The Sparrowhawk is a bird which unless observed soaring, rarely remains in view for more than a few seconds. The short rounded wings and long tail give it a distinctive silhouette compared with the

Merlin

Kestrel which has longer pointed wings, but from some angles in normal flight these two common raptors can be confused. When the hawk is observed dashing low across a field however, it is hard to mistake the two species. The Sparrowhawk is a bold hunter as it twists and turns through the trees or stalks along hedgerows, wings beating rapidly alternating with gliding, ready to pounce on any finch or sparrow which crosses its path. The much bigger and more powerful female, weighing twice as much as her partner, is able to take larger prey such as Starlings, Mistle Thrushes and even Wood Pigeons. Her bold white eye stripe and yellow eye confer a fierce countenance, and her size and sombre brown plumage readily distinguish her from the male which has slate grey back and wings and brick-red barred underparts.

During the 1950s and 1960s this species almost disappeared from those areas of Britain where crops had been regularly sprayed with the pesticide dieldrin. Sparrows, finches, pigeons and other species which fed on the grain, stored relatively innocuous doses in their tissues. When the hawks killed and fed upon these contaminated birds, a sufficiently toxic dose accumulated in their bodies. Eggs failed to hatch or broke due to the shells becoming thin and fragile. In Wales the Sparrowhawk fared better than in grain growing areas and numbers held up reasonably well. Today it is our second most common bird of prey, although in most parts you are likely to see the Buzzard far more often during a walk in the country. This is partly as a result of the Sparrowhawk's smaller size and easy concealment in its woodland habitat, and partly because its mode of hunting depending upon stealth and surprise, keeps it from view. The Sparrowhawk flies low, suddenly appearing over a hedge, turning tightly overhead before it decides in which direction to track down its prey. Sometimes, especially in March or April, it will soar alongside Buzzards over its nesting woods, or dive steeply at speed into the forest. This hawk breeds in both deciduous and conifer woods where it has benefited from the establishment of mature plantations.

All four British falcons, the Peregrine, Kestrel, Merlin and Hobby, are to be found in Wales. They are characterised by long pointed wings and shorter tails than the accipiter hawks and their flight is fast. The largest, most powerful, and in recent times the most successful of these falcons in Wales is the Peregrine. Like the Sparrowhawk, as a predominantly bird predator it suffered badly from dieldrin-based toxins in the 1950s and 1960s and disappeared from many of its cliff eyries in Wales. The last 30 years have seen a huge turn around in the fortunes of this magnificent bird of prey, which is now breeding on numerous inland crags and coastal cliffs throughout Wales. When the traditional steep crags are already occupied, Peregrines may have to accept lower, more easily accessible cliffs where unfortunately their nests are often robbed. Near my home, a riverside gorge held an eyrie which was repeatedly robbed, so that the Peregrines stopped breeding there and have not done so for four years. More recently I was surprised to see a female sitting, in full view of passing cars, on an even more vulnerable site in a sheltered alcove on a bank just above a small river. Needless to say, this nest was soon robbed.

Despite such set backs, the advance of the Peregrine has been inexorable. There are now about 350 pairs of them in Wales, many in the west and north where the mountainous country provides plenty of suitable cliff nesting sites.

Pembrokeshire, with a hundred miles of magnificent steep cliffs has a large number of Peregrines. In the gentler, lower areas of east Wales, eyries are fewer, and old disused or even worked quarries may be chosen. These are perfectly acceptable to Peregrines providing disturbance is at a minimum. One pair chose to nest in a tree in Mid Wales and more will probably do so if numbers increase further. Another pair nested on the electricity cooling tower near Swansea, an interesting choice, and one used in the total absence of cliffs in some English Midland counties in recent years.

The increase in Peregrines, like that of the Goshawk, has not been to everyone's liking. Domestic Pigeons are a favourite diet on the menu of some pairs and pigeon fanciers have sometimes taken retribution against local eyries. A threat exists also from those with an opposite motive, from those who admire the prowess of the Peregrine as a hunter. Unscrupulous collectors from far afield raid Peregrine eyries to supply a demand for falconry birds in the Middle East, where a young Peregrine may be worth thousands of pounds. Eggs are taken as well as young, and placed in incubators to prevent them becoming sterile during transit to the continent.

Watch the Peregrine circling overhead, and plunging earthwards in a breathtakingly steep dive or stoop, and it is not difficult to understand the demand for this prize species in falconry. Witness the strong winnowing wing beats, the effortless glide on angular, scythe-shaped wings, the shallow sliding movement as the Peregrine gathers momentum until it reaches incredible pace. It zooms upwards almost vertically like a jet plane, steadying itself with rapid wing beats as it reaches the top of its climb, and circles above the brow of the cliff. You stare in disbelief, wondering how any structure of flesh, blood and feathers can move so fast. This falcon has specially adapted nostrils which close in rapid flight to prevent damage to the lungs. The Peregrine is believed to attain the fastest speed of all birds, an incredible 200miles per hour In order to see a Peregrine in summer take a long coastal walk in a favoured area, or find a steep suitable cliff. No need to go too close and disturb the birds at their cliff eyrie. Many pairs can be observed with binoculars from the roadside but it is better not to attract the attention of those who might regard the falcon with mal intent. Look and listen carefully. You may hear the Peregrine before you see it; a long drawn out rasping or screaming note, echoing round the cliffs. Then a robust falcon with black hood and slate grey wings appears above the cliff. Wings beating strongly it flies round excitedly and lands on a pinnacle of rock. The nest is likely to be on a broad ledge often protected by an overhang, or sometimes it is in a disused Raven's nest, especially if the cliff provides no alternative suitable ledges. White or green splashes of excreta down the side of the rock may point either to the eyrie or to a perch or feeding post. In late May to July, you may be lucky enough to see two or three fluffy white youngsters on the nest platform. During the fledging period they turn greyer and begin to venture from the ledge before taking to the air. The parent birds - the female is larger and greyer than the male - may bring in a Pigeon, a Mistle Thrush or even a rabbit to feed the appetites of the voracious young fledglings.

Unlike the Peregrine, the Kestrel seems to have declined in the past few years in Wales. Sometimes you find it breeding on the same cliff outcrops as the Peregrine, perhaps only 100 or 200 metres away, but it may fall foul of the larger species. I can recall cliffs where 20 years ago you would see only the Kestrel, and today you see only the Peregrine. On the other hand this species is faring well at present along western cliffs which it shares with a high population of Peregrines, so its disappearance in some places may just reflect the changing fortunes of these two species. In most of Wales the Kestrel is to be found either on the coast, nesting on cliff ledges, or adjacent to the uplands where it may choose holes in trees or platforms on cliff or quarry faces to lay its five or six reddish mottled eggs.

In many parts of middle England the Kestrel is to be found on farmland habitat where it hunts for its rodent prey, hovering with unique skill over fields and roadside verges. The sharp eyes can detect the slightest twitch of any unsuspecting mouse or shrew. In my part of Wales the Kestrel is scarce in this type of improved farmland and rarely does it lay its eggs in the refurbished hedgerow nest of a Crow or similar species as it does in eastern lowlands. In south-east Wales though the situation is different. Kestrels are to be seen commonly hovering above the M4 or searching for prey over other roadside fields in Glamorgan, just as they do in counties farther to the east. Overall in Britain, the Kestrel is still the commonest bird of prey followed by Sparrowhawk and Buzzard. In Wales of course, the Buzzard holds that distinction.

Our next species, the Merlin, is a very difficult bird to track down. It is a stocky, dashing little falcon that feeds mainly on small birds which it hunts down by persistent pursuit, twisting and turning until its quarry is exhausted. On the moors in summer, Meadow Pipits, Chaffinches, Larks and other passerines are its target. In winter I have seen it chasing Starlings high above the estuary. Some years ago there was much concern about the apparent decline in numbers of this species caused by the encroachment of conifer plantations into its moorland habitat. In North Wales the Merlin usually nested on the ground amongst heather, whereas in central and South Wales the usual site was an old Crow's nest in some gnarled hawthorn on the moors. It was some time before ornithologists realised the Merlin had changed its nesting habits in South and Mid Wales, though it continues to nest in the heather farther north. Nests were found in conifers, sometimes in quite thick plantations at the edge of the moors. We were surprised to discover one ourselves just 30 metres from the side of a mountain road, 15 feet up in a spruce tree, in 1985. Such nests are not easy to find and the Merlin may often nest undetected, since it does not choose the same site each year, but pairs usually remain faithful to the same part of the moor.

The Merlin has adapted well to the intrusion of the plantations into its traditional habitat but it still requires open moorland for hunting, and blanket afforestation would present a serious threat to its future. Fortunately in Wales the evergreen forests have been developed more on a patchwork basis than in other parts of Britain, allowing a diversity of habitat which is beneficial to this and other species. Although the picture is not so bleak as originally feared, check visits to traditional sites occupied in previous years all too often find them abandoned, suggesting a decline in population of this, our smallest falcon. A few years ago I watched a pair displaying, tumbling and circling in courtship frolic over a moor near Tregaron. Returning with anticipation the

Hobby

following spring, I fully expected to be treated to the same spectacle but to my disappointment there was no sign of the Merlins. Neither has there been since. The current Welsh population of this fascinating species may be fewer than 60 pairs.

The female Merlin is not always easy to distinguish from the Kestrel. When perched on a post she appears whitish underneath and plain olive brown above, the primaries being darker. She may hover over the ground but her skills never reach the delicate wing quivering perfection of the Kestrel. Look for the mottled warm chestnut coloured back and wings of the female Kestrel, which also has a longer tail. The male Merlin is small and slate blue above, with dark streaks on dull red chest and flanks (not bars as in the Sparrowhawk), which gives him an appearance close in colour and size if not in shape to the male Sparrowhawk. When perched, the longer legs of the hawks compared with falcons may be noticeable. The Merlin is much more easily located on marshes and estuaries during wintertime than on moors in summer.

Last but not least among the falcons is the Hobby, a dark, sleek, long winged agile bird capable of catching martins and even Swifts in flight. In fact the outline of this falcon can look remarkably similar to a huge Swift as it twists and turns in the sky, grasping large insect in its talons. The Hobby feeds on insects for most of the year, but during the summer captures birds such as Swallows or Larks to feed to its young.

In England the Hobby has increased in amazing fashion from the static number of 70 or 80 pairs breeding in southern England for the past century or more, to a current total of at least 900 pairs. It now ranges into northern England, with strongholds in the East Midlands as well as the traditional haunts in the south-east and New Forest. A drive through the New Forest near dusk, when the air is thick with moths and dragonflies, may produce a dozen sightings of Hobbies. At other times the birds are easily missed, especially once incubation begins in mid June. Most Hobbies outside their nesting areas are encountered either early in May, or in August or September. Reservoirs, marshes and other places where flocks of Swallows, martins, and insect prey swarm serve as magnets to this falcon, but a bird could turn up anywhere. On warm summer evenings it is worth looking skywards to where the hirundines and Swifts gather in flocks over villages for signs of this falcon. Your attention may be drawn to a bird noticeably larger than the others and with a longer, flat-ended tail, flying with an eye-catching manoeuvrability.

The Hobby is now seen regularly over most counties of Wales. Four years ago I observed three birds over my home near Aberystwyth in August, but unfortunately a heavy shower of rain accompanied by a sudden veil of mist obscured the Hobbies from view. When it cleared a few minutes later, they had gone. On another occasion, one flew in front of my car as I approached Abergavenny, no doubt heading for the flocks of martins which were gathered over the town. In late May this year I was surprised to see one near Llanbrynmair Forest, close to the border with Meirionnydd. The bird suddenly 'appeared from nowhere' as I descended a hillside to investigate the pumping notes of a Snipe which were emanating from a small marsh. The male Hobby flew down the side of the hill twisting, turning and gliding like a Swift, crossed the marsh and made a playful pass at a startled Meadow Pipit as it flew out of sight towards the east. I was close enough to see clearly the black head and contrasting white cheek, the moustachial mark and even the rufous leg coverts.

This breeding season sighting well to the west of its normal range reminded me that the Hobby may already be nesting quite widely in Wales. Hobbies are easily overlooked, especially when they nest in hedgerows or woodlands as they usually do in Wales and the Midlands. They are more easily located in traditional clumps of scots pine on heaths and downs in the south of England. At least a dozen pairs are known to breed regularly in Wales, mainly in Gwent, with a few pairs in other counties such as Brecon and Radnor.

There are three species of harrier in Britain but only the Hen Harrier breeds regularly in Wales, about 25 pairs on the deep heather moors of the Berwyn mountains in the north east. This outpost is now the only breeding enclave for this species in southern Britain, although there are some in southern Ireland. There is deep controversy at the moment as to whether the species causes serious damage to grouse stocks in the game-keepered moors of northern Scotland, which is the stronghold of this species in Britain. Even our Welsh Harriers are under threat with a number of nests being deliberately destroyed in recent seasons by human intervention. Since the Hen Harrier nests in deep heather it is also vulnerable to predation by foxes.

The Hen Harrier is most easily observed in its winter quarters on coastal and inland marshes and estuaries, while others choose damp lower lying moors. Someone told me recently about a 'very grey hawk' he had seen on a stretch of

114

moor near Ammanford that had wintered there for the past two or three seasons and which, on investigation, turned out to be this species. The male Hen Harrier vies with the Kite for the title of 'most attractive bird of prey'. His soft grey plumage, with whitish underparts and matching black wing primaries, is a delight to behold as he quarters the bracken and heather nesting moors, or hovers above the amber and rusty winter vegetation of the marsh. The larger female may seem Buzzard-like at first, since her plumage is streaked and dark brown. But notice how she flies close to the ground, gliding on long wings held in a slight V shape, her long tail and slimmer build giving a buoyancy absent in the Buzzard. Sometimes she twists and dives headlong into the rushes, pouncing on a rodent or other prey, in a similar way to the Barn Owl. If you get a close view, the harrier's disc-like facial appearance is also reminiscent of the owls. A rear view of the bird enables you to identify her easily. Look for the diagnostic white rump, a feature shared by the very similar but slightly smaller female Montagu's Harrier. These two female harriers are sometimes difficult to distinguish in the field, and they are often then simply recorded as 'ringtails'. Season is probably the best guide away from the normal breeding areas. Any female seen between, say, October and April, is almost certain to be a Hen Harrier while the 'Montagu's' is a scarce non-breeding migrant to Wales, most often seen in late May or June. Wintering Hen Harriers can be observed regularly at Cors Caron and Cors Fochno in Ceredigion, at Llanrhidian in Glamorgan, Dowrog in Pembroke and at various sites on Anglesey. They leave their wintering quarters about mid April to return to their nesting moors.

The Montagu's Harrier is now seen but rarely in Wales, usually during the late spring. This species was never common in Britain, and now nests sparingly under the watchful eye of the RSPB in East Anglian cornfields and on the heaths of southern England, where, altogether, there are no more than just a few pairs in most years. It looks rather like a small Hen Harrier, but the male is much darker blue-grey above and has a long narrow black bar on the mid part of the wing. Unlike the male Hen Harrier, he has only a faint white rump. The main strongholds until 30 years ago were the south west moors of Devon and Cornwall, but Montagu's Harriers have long gone from those areas. In Wales, Montagu's Harriers have bred in the past at such places as Cors Caron in Ceredigion, the heaths of Dowrog near St. Davids in Pembrokeshire and the dune warrens at Newborough on Anglesey. Perhaps they will do so again, since this southern harrier may benefit from the trend towards warmer summers.

The Marsh Harrier is recorded regularly in Wales in most months of the year. This is our largest harrier and is discussed in Chapter Six. As its name suggests, it is most likely to be seen on extensive marshes and there are encouraging signs that the species will breed again in Wales in the near future. This harrier places its nest of flattened reeds deep in rank marshland vegetation which means that the species can probably never become anything but scarce as a breeding bird in this country.

Like the Hobby and the Montagu's Harrier, our next species may also benefit from warmer summers, since it feeds primarily on the larvae of wasps, and in persistent bad weather often leaves its nesting areas without attempting to breed. Also like the Hobby its headquarters used to be the New Forest,

Marsh Harrier (female)

but it has increased in numbers, though there are still no more than about 40 pairs in Britain. Since 1991 one or two pairs of Honey Buzzard have bred in Wales. No details of the area have been made public and it is a closely guarded secret, but Honey Buzzards like substantial stands of sheltered dry woodland with open glades, heath or cleared-felled areas for feeding. Up to date information puts the current population of Honey Buzzards at no less than 15 pairs in Wales so it is now well worth looking carefully at buzzard-like birds over Welsh skies.

If you see a Honey Buzzard for the first time you may have trouble distinguishing it from the common Buzzard. From underneath in good light, the striped wing pattern is noticeably different and there are three broad dark bands across the tail. The bird has a smaller head, rather slimmer body and less heavy flight. It is, in fact, more closely related to the Kite than the Buteos (Buzzards). You would need to be really lucky to see the displaying male clap his wings above his head like some gymnast, in courtship flight above a Welsh woodland.

Other species of raptor in Wales are, with the exception of the Osprey, very rare indeed. The Golden Eagle and the re-introduced White Tailed Eagle are merely occasional stragglers from their home territories in Scotland, while the Rough-legged Buzzard and rare Red-footed Falcon are mostly recorded in eastern districts of Britain. The Black Kite is just as infrequent, but we can expect to see more of this species in coming decades. The most numerous raptor in the world, found as far afield as India and Australia and as close as northern France, the Black Kite is occurring much more often in southern Britain in recent years and may well breed fairly soon. This species is smaller than the

Red Kite with a less deeply indented fork in the tail. The plumage is much darker, lacking the rich rufous tones and dark and light contrasts of the more familiar bird.

The Osprey doesn't breed in Wales though fortunate observers may watch this splendid bird dive to snatch a fish from the surface of a lake during its northward migration in April or May, or on its return to West Africa between August and October. Ospreys may turn up at traditional lakes and estuaries quite punctually as though regulated by a timing device. After the species became extinct in Scotland about 1916 they were observed less often on migration. The Osprey's returned to breed on Loch Garten in 1954 is well known and there are now 130 pairs breeding on Scottish lochs and rivers. The Scottish birds now augment individuals from Scandinavia passing twice yearly through England and Wales.

A short while ago, I was walking along a footpath near the River Severn and noticed a bird of prey half hidden behind the branches of a dead uprooted ash tree which was laying prostrate in the middle of a barley field. I could see at once from the rich uniform-brown plumage that my bird was an Osprey. With 'bated breath' I moved carefully to the right to complete the 'jigsaw' and bring the whole bird into view. The Osprey was sitting motionless with head erect, its bluish beak, white underparts and dark band through the eye now clearly visible. A local farmer later told me that an Osprey, presumably the same bird, had stayed in the vicinity for three weeks the previous June. A pair stayed for a while in spring in more traditional lakeland habitat on Llyn Clywedog in 1997, raising hopes that they might breed in Wales. Perhaps one day they will!

BIRD OF PREY - species to look for:

Buzzard	Hen Harrier
Sparrowhawk	Marsh Harrier
Kestrel	Hobby
Peregrine	Osprey
Red Kite	Montagu's Harrier
Goshawk	Honey Buzzard
Merlin	

Notes: Generally, species are progressively harder to find as you move through the list from Buzzard and Sparrowhawk to Montagu's Harrier and Honey Buzzard. All other raptor species are vagrants in Wales.

Osprey

SYSTEMATIC LIST OF WELSH BIRDS

All birds regularly observed in Wales are included in this section. Some of the more frequently recorded vagrants are entered separately at the end of this list. Welsh and Latin (scientific) names are in italics.The relative abundance or rarity of breeding birds (breeding status rating =BR) is denoted by a scale, one to seven, the higher number indicating greater rarity.

RED-THROATED DIVER   *Trochydd Gyddfgoch   Gavia stellata*
Winter visitor and passage migrant offshore around all coasts, especially north and mid Cardigan Bay and Carmarthen Bay.

BLACK-THROATED DIVER   *Trochydd Gyddfddu   G.arctica*
Very scarce visitor in winter and on passage, mostly Glamorgan and Pembrokeshire.

GREAT NORTHERN DIVER   *Trochydd Mawr   G. immer*
More frequent than the last species, small numbers winter round the Welsh coast or occur on passage. Unlike the other divers, sometimes visits inland waters.

LITTLE GREBE   *Gwyach Fach   Tachybaptus ruficollis*
Resident and winter visitor to lakes and pools with adequate vegetation. BR4

GREAT CRESTED GREBE   *Gwyach Fawr Gopog   Podiceps cristatus*
Local resident to suitable lakes, notably on Anglesey and eastern counties. Scarce or absent in many western areas. More common as a winter visitor, on freshwater or the sea, with large annual concentrations in Cardigan Bay. BR5

RED-NECKED GREBE   *Gwyach Yddfgoch   P. grisegena*
Scarce winter visitor to coastal waters, no more than a handful of records most years.

SLAVONIAN GREBE   *Gwyach Gorniog   P. auritus*
Scarce visitor in winter, usually in rather larger numbers than the last species.

BLACK-NECKED GREBE   *Gwyach Yddfddu   P. nigricollis*
Rare spring and late summer migrant or winter visitor. Formerly bred.

FULMAR   *Aderyn-Drycin y Graig   Fulmarus glacialis*
Breeds on suitable cliffs chiefly in western and northern counties from Pembrokeshire to Clwyd.
Also smaller numbers Glamorgan. BR3

SOOTY SHEARWATER   *Aderyn-Drycin Du   Puffinus griseus*
Regularly recorded each year on migration routes off western coasts and Pembrokeshire islands.

BALEARIC (MEDITERRANEAN) SHEARWATER
Aderyn-Drycin Mor y Canoldir   P. yelkouan
Scarce but regular visitor, mainly seen near islands and headlands in late summer and autumn.

MANX SHEARWATER   *Aderyn-Drycin Manaw   P. puffinus*
Large breeding colonies on Skomer, Skokholm, Ramsey and Bardsey. Elsewhere in favoured locations, feeding parties or passage movements involving 100s or even 1000s of birds occur in summer and autumn. BR4

STORM PETREL   *Pedryn Drycin   Hydrobates pelagicus*
Breeds on the Pembrokeshire islands of Skokholm (up to 8000 pairs), Skomer and Ramsey. Summer movements occur along Glamorgan coast and small numbers seen offshore at other localities, especially after westerly gales. BR5

LEACH'S PETREL   *Pedryn Gynffon-Fforchog   Oceanodroma leucorrhoa*
Scarce but regular autumn passage migrant, mainly to headlands, islands and the Dee estuary.

GANNET   *Hugan   Morus bassanus*
Huge colony of 30,000 pairs breed on Grassholm. Commonly observed off-shore at most times of the year around the Welsh coast. BR4

CORMORANT   *Mulfran   Phalacrocorax carbo*
Among our commonest seabirds, the Cormorant is resident all year in bays, harbours, and all kinds of marine habitats. Also frequently inland on lakes and reservoirs. Nests colonially on cliffs. Biggest colony is on St Margaret's Isle, Tenby. BR2

SHAG   *Mulfran Werdd   P. aristotelis*
Resident, breeds more thinly than the last species, two or three pairs some-times breeding among auks and other sea birds. Occurs on western cliffs from Great Orme to Gower. BR4

BITTERN   *Aderyn y Bwn   Botaurus stellaris*
Rare visitor, usually in winter. Has bred Anglesey.

LITTLE EGRET   *Creyr Bach   Egretta garzetta*
Recorded with increasing numbers and regularity, mainly at coastal marshes in South Wales. The majority of birds are seen in winter and fewest are re-corded in June and July. Perhaps 50-100 birds are involved in total, and breed-ing may soon occur. Ten years ago this species was only a vagrant in Wales.

GREY HERON   *Creyr Glas   Ardea cinerea*
Resident species. Nests colonially in trees at sites scattered throughout Wales. BR3

SPOONBILL   *Llwybig   Platalea leucorodia*
Scarce or rare visitor, in summer or autumn usually, occasionally in winter. Recorded annually and appears to be increasing.

MUTE SWAN   *Alarch Dof   Cygnus olor*
Resident, breeding locally in Wales, scarcer in mid Wales and upland districts. Commoner in the south, north-west and eastern most areas. Many non-breed-ing birds on lakes, rivers and estuaries. BR3

BEWICK'S SWAN   *Alarch Bewick   C. columbianus*
Regular but local winter visitor, mainly to Gwent, Radnor, Brecon, Carmarthenshire, and parts of north-west Wales.

WHOOPER SWAN   *Alarch y Gogledd   C. cygnus*
Regular winter visitor in rather small numbers. Main areas are Glaslyn marshes and Anglesey in the north, Cors Caron in Ceredigion, and in the Severn and Wye valleys in eastern Mid Wales.

PINK-FOOTED GOOSE   *Gwydd Droed-Binc   Anser brachyrhynchus*
Scarce winter visitor, mainly to North Wales.

WHITE-FRONTED GOOSE   *Gwydd Dalcen-Wen   A. albifrons*
Very local and declining winter visitor. Main flocks occur Dyfi estuary (Greenland race) and Anglesey (Russian race). Flock at Dryslwyn, Carmarthenshire seems to have gone now.

GREY-LAG GOOSE   *Gwydd Wyllt   A. anser*
Feral flocks occur in suitable habitats and some pairs breed. In the wild this goose is a rare winter visitor.

CANADA GOOSE   *Gwydd Canada   Branta canadensis*
This introduced species is increasing in many parts of Wales. Large flocks occur in places such as the Dee, Dyfi, and Cleddau estuaries, and on Llangorse lake. BR2

BARNACLE GOOSE   *Gwydd Wryan   B. leucopsis*
Scarce and very local. Currently a few winter on the Dyfi and at Marloes in Pembrokeshire.

BRENT GOOSE   *Gwydd Ddu   B. bernicla*
Up to 1000 birds winter on the Burry Inlet in Glamorgan. Elsewhere a few are found at various suitable locations round the Welsh coast.

SHELDUCK   *Hwyaden yr Eithin   Tadorna tadorna*
Breeds commonly on suitable low-lying coasts and estuaries. Large gatherings can be seen in autumn and winter at the Rhymney estuary, Burry Inlet, Cleddau, and Dyfi, with smaller numbers in other estuaries. BR3

WIGEON   *Chwiwell   Anas penelope*
Locally common winter visitor, mainly coastal to lakes, marshes and estuaries. Scarcer inland. Sometimes breeds in North Wales.

GADWALL   *Hwyaden Lwyd   A. strepera*
Breeds on Anglesey, otherwise winter visitor in small numbers.BR6

TEAL   *Corhwyaden   A. crecca*
Declining breeding bird to lakes and upland pools. One of the most common species of duck in winter. BR5

MALLARD   *Hwyaden Wyllt   A. platyrhynchos*
Abundant and widespread, both as a breeding bird and as a winter visitor in all kinds of watery habitats. BR1

PINTAIL   *Hwyaden Lostfain   A. acuta*
Fairly common winter visitor, mainly to estuarine habitats. Has bred.

GARGANEY   *Hwyaden Addfain   A. querguedula*
Scarce passage migrant in spring and autumn. Has bred occasionally.

SHOVELER   *Hwyaden Lydanbig   A. clypeata*
Scarce breeding bird, mainly Anglesey, sometimes Carmarthenshire. More frequent and not uncommon as a winter visitor. BR6

POCHARD   *Hwyaden Bengoch   Aythya ferina*
Breeds Anglesey, and more sporadically in a few other counties, notably Carmarthenshire. A fairly common winter visitor, especially to coastal freshwater lakes. BR6

RED-CRESTED POCHARD   *Hwyaden Gribgoch   Netta rufina*
Scarce visitor, but most records are of escaped feral birds.

TUFTED DUCK   *Hwyaden Gopog   Aythya fuligula*
Breeds locally in many counties, but scarce or absent in western areas except Anglesey. Common in winter. BR5

SCAUP   *Hwyaden Benddu   A. marila*
Scarce and local winter visitor to coastal waters.

EIDER   *Hwyaden Fwythblu   Somateria mollissima*
Very local winter visitor; main sites Aberdysynni and Burry Inlet. Smaller numbers remain throughout the year. Bred Puffin Island 1997.

LONG-TAILED DUCK   *Hwyaden Gynffon Hir   Clangula hyemalis*
Scarce winter visitor, sometimes staying till summer, most often to coasts of North Wales.

Eider

COMMON SCOTER   *Mor-Hwyaden Ddu*   *Melanitta nigra*
Winter visitor, some stay till summer, largest concentrations are found in
Carmarthen Bay, smaller numbers in Cardigan Bay.

VELVET SCOTER   *Mor Hwyaden y Gogledd*   *M. fusca*
Much scarcer than the last species, a few occur in the coastal waters of most
maritime counties in winter.

GOLDENEYE   *Hwyaden Lygad-Aur*   *Bucephala clangula*
Locally common in winter on lakes and some river estuaries such as the Cleddau
and Ogmore.

SMEW   *Lleian Wen*   *Mergus albellus*
Rare winter visitor, mainly to lakes in South Wales and most frequent in Glamorgan.

RED-BREASTED MERGANSER   *Hwyaden Frongoch*   *M. serrator*
Breeds usually on rivers near to the coast in north-west Wales as far south as
Ceredigion. Shows signs of expanding further south. Common in winter, and
non-breeding flocks occur in North and Mid Wales in summer. BR4

GOOSANDER   *Hwyaden Ddanheddog*   *M. merganser*
Now nests on rivers in most parts of Wales. Commoner in winter when birds
are most easily seen on lakes and reservoirs. BR4

RUDDY DUCK   *Hwyaden Goch*   *Oxyura jamaicensis*
A few pairs breed on Anglesey, and in Clwyd and Montgomery. Large winter
flocks occur on Anglesey, a few are seen elsewhere.

HONEY BUZZARD   *Bod y Mel*   *Pernis apivorus*
Rare summer visitor since about 1990. Currently about 15 pairs are breeding in
Wales. BR7

RED KITE   *Barcud*   *Milvus milvus*
Resident in Wales but some move southwards in autumn. Present breeding
population may be as high as 150 pairs. BR5

MARSH HARRIER   *Bod y Gwerni*   *Circus aeruginosus*
Scarce migrant or winter visitor. Sometimes one or two stay over summer and
this species may nest again, as it has done in the past.

HEN HARRIER   *Bod Tinwen*   *C. cyaneus*
Between 20 and 30 pairs breed in north-east Wales. More birds join the residents to winter on coastal or inland marshes or sheltered moors. BR6

MONTAGU'S HARRIER   *Bod Montagu*   *C. pygargus*
Occasional summer migrant, formerly bred.

GOSHAWK   *Gwalch Marth*   *Accipiter gentilis*
Uncommon but increasing resident, rising from nothing 15 years ago to a current total of perhaps 300 pairs. BR4

122

SPARROWHAWK   *Gwalch Glas   A nisus*
Common resident in conifer and deciduous woodlands. BR2

BUZZARD   *Bwncath   Buteo buteo*
Common and conspicuous resident throughout most of Wales. Rather less plentiful in the north and few on Anglesey. BR2

OSPREY   *Gwalch y Pysgod   Pandion haliaetus*
Regular visitor to lakes and estuaries on spring and autumn passage in very small numbers.

KESTREL   *Cudyll Coch   Falco tinnunculus*
Common resident in most parts of Wales, particularly to coastal cliffs, rough pasture and uplands. May have declined recently. B3

MERLIN   *Cudyll Bach   F. columbarius*
Scarce resident, breeding on heather moors in the north and at the edge of plantations bordering moorland further south. May be declining. BR6

HOBBY   *Hebog yr Ehedydd   F. subbuteo*
Rare breeding summer visitor to Gwent, Radnor, Brecon and possibly other eastern counties. Numbers are increasing. BR7

PEREGRINE   *Hebog Tramor   F. peregrinus*
About 300 pairs breed on coastal and inland cliffs. Commonest in Pemrokeshire Caernarfonshire and central Cambrian mountains. BR4

RED GROUSE   *Grugiar   Lagopus lagopus*
Resident on upland heather moors, mainly now in north-east. This species is in decline due to loss of habitat, and there are now merely residual populations in many former haunts. BR4-5

BLACK GROUSE   *Grugiar Ddu   Tetrao tetrix*
This species is in serious decline. The main concentration is in Meirionydd, Clwyd and Montgomery, while a few birds can still be found in north Ceredigion. Elsewhere the remaining breeding birds have gone. BR6

RED-LEGGED PARTRIDGE   *Petrisen Goesgoch   Alectoris rufa*
Very scarce resident to easternmost parts of Wales and Anglesey. BR6

GREY PARTRIDGE   *Petrisen   Perdix perdix*
A few pairs breed in Anglesey and eastern counties, notably Gwent. Formerly common. BR6

QUAIL   *Sofliar   Coturnix coturnix*
A rare summer migrant, breeding spasmodically, especially during warm summers with prevalent south-easterly winds. BR7

PHEASANT   *Ffesant   Phasianus colchicus*
Common resident, especially in low lying districts. Local in distribution but widespread throughout the country. BR2

WATER RAIL   *Rhegen y Dwr*   *Rallus aquaticus*
Scarce resident breeding species at widely separated marshes throughout Wales. Easily overlooked. Also winter visitor. BR6

CORNCRAKE   *Rhegen yr Yd*   *Crex crex*
Rare migrant. Formerly common in hay meadows. One or two recent breeding records have not been sustained.

SPOTTED CRAKE   *Rhegen Fraith*   *Porzana porzana*
Rare, recorded most years. Easily overlooked, has bred and could do so periodically.

MOORHEN   *Iar Ddwr*   *Gallinula chloropus*
Common resident but surprisingly local in upland areas with acidic pools and lakes. More numerous in winter. BR3

COOT   *Cwtiar*   *Fulica atra*
Familiar and common breeding bird on lakes and marshy pools, and numbers augmented by winter visitors. BR2

OYSTERCATCHER   *Pioden y Mor*   *Haematopus ostralegus*
Nests thinly on suitable beaches all round the Welsh coast. Large gatherings occur on main estuaries in winter. BR3

AVOCET   *Cambig*   *Recurvirostra avosetta*
Rare on passage, most often South Wales.

LITTLE RINGED PLOVER   *Cwtiad Torchog Bach*   *Charadrius dubius*
Scarce summer visitor and passage migrant, breeding on shingle banks of rivers such as the Twyi, Severn, Wye, Usk and Conwy. May soon colonise other rivers. BR6

RINGED PLOVER   *Cwtiad Torchog*   *C. hiaticula*
Resident breeder, rather scarce on suitable sand and shingle beaches in parts of Wales. Common winter visitor and passage migrant. BR4

DOTTEREL   *Hutan y Mynydd*   *Eudromias morinellus*
Scarce and local, but regular passage migrant. Has bred.

GOLDEN PLOVER   *Cwtiad Aur*   *Pluvialis apricaria*
Nests in decreasing numbers, mainly in the Elenydd region of Mid Wales, with some pairs still breeding in the Mynydd Hiraethog area of Clwyd, and a few other districts. Locally common winter visitor, especially to coastal districts. BR6

GREY PLOVER   *Cwtiad Llwyd*   *P. squatarola*
A fairly common winter visitor to suitable estuarine areas, generally scarcer along western coasts.

LAPWING   *Cornchwiglen*   *Vanellus vanellus*
This once familiar species is now a scarce breeding bird in most of Wales. In winter it is more common occurring in flocks on fields, flooded meadows, salt marsh and beaches. BR4

KNOT *Pibydd yr Aber* *Calidris canutus*
Local winter visitor but occurs in huge flocks in its favoured haunts, notably on the Burry Inlet and the Dee estuary.

SANDERLING *Pibydd y Tywod* *C. alba*
Uncommon winter visitor, most frequently Carmarthen and Glamorgan. More numerous as a passage migrant.

LITTLE STINT *Pibydd Bach* *C. minuta*
A rather uncommon passage migrant in variable numbers, often to coastal fresh water pools or brackish marsh. September is the best month usually.

CURLEW SANDPIPER *Pibydd Cambig* *C. ferruginea*
Scarce passage migrant, occurs at similar times and places to the last species and is also more frequent in autumn than in spring.

PURPLE SANDPIPER *Pibydd Du* *C. maritima*
Local winter visitor to rocky shores from October till May.

DUNLIN *Pibydd y Mawn* *C. alpina*
Numerous winter visitor and passage migrant. Breeds in small numbers mainly in Elenydd district of Mid Wales and a few places in the north. BR6

RUFF *Pibydd Torchog* *Philomachus pugnax*
Scarce passage migrant and occasional winter visitor, this species may occur in any month of the year. Has bred Anglesey.

JACK SNIPE *Giach Fach* *Lymnocryptes minimus*
Scarce and easily overlooked winter visitor to freshwater marsh or estuarine spartina.

SNIPE *Giach Gyffredin* *Gallinago gallinago*
Declining resident breeding bird and more commonly, a winter visitor. Wet upland areas are generally favoured for nesting. BR4

WOODCOCK *Cyffylog* *Scolopax rusticola*
Thinly scattered breeding bird, in damp conifer or deciduous woods. Particularly scarce in western parts of Wales. Commoner in winter. BR5

BLACK-TAILED GODWIT *Rhostog Gynffonddu* *Limosa limosa*
Local passage migrant, restricted to just a few locations in winter, notably Penclacwydd in Carmarthen and the Dee estuary which attracts totals of nearly 2000 birds, a figure of national importance in the UK.

BAR-TAILED GODWIT *Rhostog Gynffonfrith* *L. lapponica*
Local passage migrant and winter visitor to coastal districts. Main site is the Burry Inlet, otherwise most frequent in parts of South Wales and Inland Sea (Anglesey) and Dee estuary.

WHIMBREL *Coegylfinir* *Numenius phaeopus*
Regular passage migrant, more common in spring than autumn. Mainly seen on river estuaries.

CURLEW   *Gylfinir   N. arquata*
Breeds on wet pastures, grass and heather moors and marginal land close to
the uplands. Declining. Common in coastal habitats in winter. BR4

SPOTTED REDSHANK   *Pibydd Coesgoch Mannog   Tringa erythropus*
Fairly scarce passage migrant to coastal districts, a few winter, mostly in Clwyd
and south-west Wales.

REDSHANK   *Pibydd Coesgoch   T. totanus*
Increasingly scarce breeding resident. Common in winter on estuaries, mud
flats and coastal marshes. BR6

GREENSHANK   *Pibydd Coeswerdd   T. nebularia*
Fairly common passage migrant, mainly to coastal counties where it usually
frequents lakes and tidal river banks. Some stay over winter.

GREEN SANDPIPER   *Pibydd Gwyrdd   T. ochropus*
Uncommon passage migrant. Small numbers winter in coastal areas, mainly
in South Wales.

WOOD SANDPIPER   *Pibydd y Graean   T. glareola*
Scarce passage migrant in small but variable numbers, usually less than 20 of
them recorded in any one year.

COMMON SANDPIPER   *Pibydd y Dorlan   Actitis hypoleucos*
A fairly common summer visitor, breeding by rivers and upland pools. Also
passage migrant and occasional winter visitor. BR4

TURNSTONE   *Cwtiad y Traeth   Arenaria interpres*
Common winter visitor to rocky shores, and a few non-breeding birds stay
during the summer.

GREY PHALAROPE   *Llydandroed Llwyd   Phalaropus fulicarius*
Chiefly a scarce autumn passage migrant to coastal districts.

POMERINE SKUA   *Sgiwen Frech   Stercorarius pomarinus*
Scarce but regular passage on migration routes, passing headlands and islands
off the Welsh coast, particularly after westerly gales.

ARCTIC SKUA   *Sgiwen y Gogledd   S. parasiticus*
Similar to the above species, but observed more frequently, and usually in larger
numbers.

LONG-TAILED SKUA   *Sgiwen Lostfain   S. longicaudus*
Passage migrant, rare.

GREAT SKUA   *Sgiwen Fawr   Catharacta skua*
Regular but uncommon passage migrant. As in the case with other skuas,
Strumble Head provides the most records. Bardsey, Point Lynas, St David's,
and headlands in Glamorgan, are also good sea-watching points for this and
other species of skua.

MEDITERRANEAN GULL   *Gwylan Mor y Canoldir*   *Larus melanocephalus*
Uncommon but regular non-breeding visitor, recorded in all months in coastal
counties and commonest in Glamorgan (notably Blackpill).

LITTLE GULL   *Gwylan Fechan*   *L. minutus*
Uncommon, but like the last species, recorded throughout most of the year.
Rarely in June and July. Most often inland. This species also does not breed in
Wales.

Great Skua and Arctic Skua (right)

BLACK-HEADED GULL   *Gwylan Benddu*   *L. ridibundus*
Common breeding bird but patchy in distribution, mainly to Anglesey and
some upland areas of North and Mid Wales. Abundant visitor, chiefly to coasts
at all times of the year. BR3

RING-BILLED GULL   *Gwylan Fodrwybig*   *L. delawarensis*
Former vagrant, now regular, particularly at Blackpill near Swansea. Best
months are usually February until May.

COMMON GULL   *Gwylan y Gweunydd*   *L. canus*
Fairly common and widespread winter visitor, especially to South Wales. Large
numbers at Blackpill and Burry Inlet where this species can be seen at all months
of the year. Formerly bred on Anglesey.

LESSER BLACK-BACKED GULL   *Gwylan Gefnddu Leiaf*   *L. fuscus*
Common but local nesting species, with large colonies on Pembrokeshire islands, Flatholm, Cardigan Island and Bardsey. Nearly 20,000 pairs on Skomer.
Nests on rooftops in the major towns in South Wales. Abundant in winter in
much of the south, scarcer elsewhere. BR3

HERRING GULL   *Gwylan y Penwaig*   *L. argentatus*
Abundant breeding resident and winter visitor. BR1

ICELAND GULL   *Gwylan yr Arctig*   *L. glaucoides*
Very scarce winter visitor, mostly to South Wales.

GLAUCOUS GULL   *Gwylan y Gogledd*   L. hyperboreus
Scarce visitor, mostly winter, but may be seen in any month.

GREAT BLACK -BACKED GULL   *Gwylan Gefnddu Fwyaf*   *L. marinus*
Widespread resident on western coasts from Anglesey and Caernarfon down
to Pembrokeshire, the largest numbers being on islands. Common in winter.
BR3

KITTIWAKE   *Gwylan Goesddu*   *Rissa tridactyla*
Summer visitor, breeding at cliff colonies in western coastal counties and also
passage migrant. BR3

SANDWICH TERN   *Morwennol Bigddu*   *Sterna sandvicensis*
Passage migrant along Welsh coasts, breeds on Anglesey. BR5

ROSEATE TERN   *Morwennol Wridog*   *S. dougallii*
Rare summer visitor to Anglesey. BR7.

COMMON TERN   *Morwennol Gyffredin*   *S. hirundo*
Summer visitor to Anglesey and Flint. Passage migrant along coasts. BR5

ARCTIC TERN   *Morwennol y Gogledd*   *S. paradisaea*
Breeds on Anglesey, observed on passage at other coastal locations. BR5

LITTLE TERN   *Morwennol Fechan*   *S. albifrons*
Breeds now at only one colony in Wales, near Point of Ayr in Flint. Small numbers elsewhere on passage. BR6

BLACK TERN   *Corswennol Ddu*   *Chlidonias niger*
A rather scarce passage migrant offshore or sometimes inland, more often autumn than spring.

GUILLEMOT   *Gwylog*   *Uria aalge*
Breeds at rock face colonies on western coasts, moving out to sea in July and
returning to the coast in December. BR4

RAZORBILL   *Llurs*   *Alca torda*
Usually occurs in rather smaller colonies than the last species but often shares
the same or adjacent cliffs. Leaves its colonies in August to moult far out at
sea, returning to the vicinity about February. Some winter around the coast.
BR4

BLACK GUILLEMOT   *Gwylog Ddu*   *Cepphus grylle*
Breeds at one location only, at Fedw Fawr on Anglesey. Rarely seen elsewhere.
BR7

128

LITTLE AUK   *Carfil Bach   Plautus alle*
Rare winter visitor, most often seen after severe gales. Regular off Strumble Head.

PUFFIN   *Pal   Fratercula arctica*
Nesting colonies large but few, notably Skomer and Skockholm. Elsewhere there are a few smaller groups mainly on islands, but also at South Stack on Anglesey. BR5

ROCK DOVE/FERAL PIGEON   *Colomen y Graig   Columba livia*
Scarce if not extinct in it's 'pure' Rock Dove form, feral pigeons breeds on many suitable cliffs. BR3

STOCK DOVE   *Colomen Wyllt   C. oenas*
Resident, commoner in areas where there is mixed farming. Scarcer in parts of west Wales. BR3

WOOD PIGEON   *Ysguthan   C. palumbus*
Abundant resident wherever there are trees. BR1

COLLARED DOVE   *Turtur Dorchog   Streptopelia decaocto*
Common resident in gardens, towns and villages. BR2

TURTLE DOVE   *Turtur   S turtur*
Scarce summer visitor and declining, now breeds only in Gwent. Elsewhere occasional passage migrant. BR7

CUCKOO   *Cog   Cuculus canorus*
Thinly distributed summer visitor to all parts of Wales. BR4

BARN OWL   *Tylluan Wen   Tyto alba*
Sparsely distributed resident to farmland in all counties. BR4

LITTLE OWL   *Tylluan Fach   Athene noctua*
Rather scarce in most parts of the country. Commonest in Gower, Lleyn, Anglesey, the north-east, and Gwent. BR5

TAWNY OWL   *Tylluan Frech   Strix aluco*
This species is much the commonest owl in Wales. BR2

LONG-EARED OWL   *Tylluan Gorniog   Asio otus*
Rare resident and winter visitor, may be overlooked. BR7

SHORT-EARED OWL   *Tylluan Glustiog   A. flammeus*
Scarce and spasmodic breeding bird, most frequently on Pembrokeshire islands. Could nest on almost any suitable moor or upland plantation. Also winter visitor in very small numbers. BR6-7

NIGHTJAR   *Troellwr Mawr   Caprimulgus europaeus*
This scarce summer visitor is once again increasing after a long period of decline. Breeds on upland plantations. BR5

SWIFT   *Gwennol Ddu   Apus apus*
Common summer visitor mainly to towns and villages .BR2

KINGFISHER   *Glas y Dorlan   Alcedo atthis*
Resident, typically on less turbulent sections of rivers with sandy or clay banks suitable for nesting. BR4

HOOPOE   *Copog   Upupa epops*
Scarce passage migrant, chiefly in spring. Bred in 1997.

WRYNECK   *Pengam   Jynx torquilla*
Very scarce migrant, mainly in autumn, often recorded from headlands and islands.

GREEN WOODPECKER   *Cnocell Werdd   Picus viridis*
Resident, common in the south and east, scarce in some western areas. BR4

GREAT SPOTTED WOODPECKER   *Cnocell Fraith Fwyaf   Dendrocopos major*
Common resident species in well wooded districts throughout Wales. BR3

LESSER SPOTTED WOODPECKER   *Cnocell Fraith Leiaf   D. minor*
Uncommon and local resident in mature or damp woodlands, found sparingly in most counties. BR5

WOODLARK   *Ehedydd y Coed   Lullula arborea*
 Former breeding species, now only recorded rarely as an autumn migrant.

SKYLARK   *Ehedydd   Alauda arvensis*
Common but declining resident, mainly on moors, commons and rough pasture. BR2

SHORE LARK   *Ehedydd y Traeth   Eremophila alpestris*
Rare winter visitor to the coast, most often in North Wales.

SAND MARTIN   *Gwennol y Glennydd   Riparia riparia*
Locally common summer visitor, most numerous in Carmarthen, Gwent, and Brecon. BR3

SWALLOW   *Gwennol   Hirundo rustica*
An abundant summer visitor throughout Wales. BR1

HOUSE MARTIN   *Gwennol y Bondo   Delichon urbic*
 A very common and widespread summer visitor. BR2

TREE PIPIT   *Corhedydd y Coed   Anthus trivialis*
Common summer visitor to wooded heaths and plantations. Absent on Anglesey. BR3

MEADOW PIPIT   *Corhedydd y Waun   A. pratensis*
Abundant resident, this is the commonest species on moorland and is found in many kinds of rough grassland habitats. BR1

ROCK PIPIT   *Corhedydd y Graig   A. petrosus*
Resident to coastal counties except Flint and Gwent. Also winter visitor to all
coasts. BR3

WATER PIPIT   *Corhedydd y Dwr   A. spinoletta*
Scarce but increasingly recorded winter visitor and passage migrant, mostly
to estuarine locations.

YELLOW WAGTAIL   *Siglen Felen   Motacilla flava*
Uncommon and local summer visitor, mainly to eastern river valleys, the Gwent
Levels, and sometimes elsewhere. Otherwise a passage migrant in small num-
bers at watery locations. BR6

GREY WAGTAIL   *Siglen Lwyd   M. cinerea*
Common resident on fast flowing streams and rivers. BR3

PIED WAGTAIL   *Siglen Fraith   M. alba*
Abundant on farms and all kinds of waterside habitats. BR1

WAXWING   *Cynffon Sidan   Bombycilla garrulus*
Rare winter visitor, flocks occurring during eruptive years.

DIPPER   *Bronwen y Dwr   Cinclus cinclus*
Resident of fast flowing rivers and streams. BR4

WREN   *Dryw   Troglodytes troglodytes*
Abundant species, adapting to all kinds of habitat from sea level to mountain
tops. BR1

DUNNOCK   *Llwyd y Gwrych   Prunella modularis*
Abundant wherever there are hedges and bushes for nesting. BR1

ROBIN   *Robin Goch   Erithacus rubecula*
Abundant and widespread resident. BR1

NIGHTINGALE   *Eos   Luscinia megarhynchos*
Formerly bred, most recently in Gwent, but now is only a scarce passage migrant in Wales.

BLACK REDSTART   *Tingoch Du   Phoenicurus ochruros*
Uncommon but regular passage migrant and winter visitor. Most frequent in
South Wales where it occasionally breeds. BR7

REDSTART   *Tingoch   P. phoenicurus*
Common summer visitor to open woodland and other habitats with at least a
scattering of trees. Scarce Anglesey and Pembroke. BR2

WHINCHAT   *Crec yr Eithin   Saxicola rubetra*
Summer visitor to upland and marginal habitats close to the hills. Less frequent Anglesey, Pembrokeshire, and some eastern districts. BR3

STONECHAT   *Clochdar y Cerrig   S. torquata*
Resident throughout the year unlike the last species, and more coastal in distribution. Usually nests in gorse, less often heather covered hill slopes. BR3

WHEATEAR   *Tinwen y Garn   Oenanthe oenanthe*
Locally common on coasts backed by stony or rocky hillsides, and similar haunts in upland areas. BR3

RING OUZEL   *Mwyalchen y Mynydd   Turdus torquatus*
This mountain species is becoming scarce in its rocky upland haunts. Most frequent in North Wales, Elan valley and a few areas further south. BR5

BLACKBIRD   *Mwyalchen   T. merula*
Abundant and familiar species wherever there are shrubs, trees and hedgerows. BR1

FIELDFARE   *Socan Eira   T. pilaris*
Very common and widespread winter visitor, typically feeding on hedgerow berries or searching for grubs in fields. Always in flocks.

SONG THRUSH   *Bronfraith   T. philomelos*
Once abundant, the much loved Song Thrush is seriously declining in its farmland and garden haunts. Common in conifer plantations. Numbers are augmented from abroad in winter. BR2

REDWING   *Coch Dan-Aden   T. iliacus*
Very common winter visitor. Often forms mixed flocks with Fieldfares.

MISTLE THRUSH   *Brych y Coed   T. viscivorus*
Abundant resident, especially in the south. Seems to like sheltered farmland adjacent to the uplands. BR2

Blackcap (left) and Garden Warbler (both males)

132

CETTI'S WARBLER   *Telor Cetti   Cettia cetti*
Scarce or rare, this species, unlike most warblers, stays in this country throughout the year. So far there are only a handful of breeding sites in Wales. BR5-6.

GRASSHOPPER WARBLER   *Troellwr Bach   Locustella naevia*
Local summer visitor to marshes, bogs and plantations. BR5

SEDGE WARBLER   *Telor yr Hesg   Acrocephalus schoenobaenus*
Locally common summer visitor to marshes, overgrown ditches and bog with
tangled vegetation. Tends to be commoner in low-lying districts. BR3

REED WARBLER   *Telor y Cyrs   A. scirpaceus*
Very local summer visitor, confined to tall beds of phragmites reeds, mostly in
coastal areas or the eastern borders of Wales. BR 4

DARTFORD WARBLER   *Telor Dartford   Sylvia undata*
Very scarce visitor from heaths of southern England, but bred in 1998 and
could colonise gorse commons in South Wales. This species is resident in Britain

LESSER WHITETHROAT   *Llwydfron Fach   S. curruca*
Scarce summer visitor to low-lying sheltered districts, this species is commoner
in the south-east and appears to be increasing in Wales. BR5

WHITETHROAT   *Llwydfron   S. communis*
Locally common, particularly in coastal areas with plenty of scrub. Shuns higher
ground except plantations with plenty of low vegetation. As with most warblers, a summer visitor. BR3

BLACKCAP   *Telor Penddu   S. atricapilla*
Abundant summer visitor to woods and copses. Some migrate here over winter from Europe. BR1

GARDEN WARBLER   *Telor yr Ardd   S. borin*
Despite its name, this summer visitor seems to be even more suited to the
upland woods than the Blackcap, provided there are ample shrubs and tangled undergrowth. BR1

WOOD WARBLER   *Telor y Coed   Phylloscopus sibilatrix*
Common in upland conifer or deciduous woods where there is sparse vegeta-
tion beneath the canopy. Scarce Anglesey and parts of Pembrokeshire and south
Glamorgan. Summer visitor. BR3

CHIFFCHAFF   *Siff-Saff   Ph. collybita*
Very common summer visitor to sheltered woods with thick undergrowth.
Some stay the winter. BR2

WILLOW WARBLER   *Telor yr Helyg   Ph. trochilus*
Abundant summer visitor, preferring more open ground than the previous
species. BR1

133

GOLDCREST   *Dryw Eurben   Regulus regulus*
Abundant in conifer woods, and even gardens and churchyards where there are suitable trees. Much scarcer after hard winters.BR1

FIRECREST   *Dryw Penfflamgoch   Regulus ignicapillus*
Scarce passage migrant. Bred for almost a decade in both Montgomery and Gwent until about 1990, and could easily do so again. BR7

SPOTTED FLYCATCHER   *Gwybedog Mannog   Muscicapa striata*
Fairly common summer visitor but may have declined lately. BR3

PIED FLYCATCHER   *Gwybedog Brith   Ficedula hypoleuca*
Common summer visitor to upland deciduous woods throughout most of Wales. Scarcer in parts of the far south and south-west, and rare on Anglesey and Lleyn. BR3

BEARDED TIT   *Titw Barfog   Panurus biarmicus*
A scarce visitor to Wales, but bred for a while in Glamorgan and has bred in Anglesey, so it may well do so again. BR7

LONG-TAILED TIT   *Titw Gynffon-Hir   Aegithalos caudatus*
Very common resident in hedgerows, open woodland, scrub and copses. More numerous in the south, local on Anglesey. BR2

MARSH TIT   *Titw'r Wern   Parus palustris*
Locally common in deciduous woodland, scarce or absent in the north and west between Ceredigion and Anglesey. BR4

WILLOW TIT   *Titw'r Helyg   P. montanus*
Locally common. Like the last species, more frequent in the south and east of the country. Prefers damper more open ground. BR3

COAL TIT   *Titw Penddu   P. ater*
Numerous species, chiefly but not exclusively in coniferous woodland. BR2

BLUE TIT   *Titw Tomas Las   P. caeruleus*
Abundant almost everywhere in suitable habitat. BR1

GREAT TIT   *Titw Mawr   P. major*
Almost as numerous as the Blue Tit, both species greatly benefit from winter feeding and the provision of nest boxes. BR1

NUTHATCH   *Delor y Cnau   Sitta europaea*
Common and widespread resident in deciduous woodland. Less numerous in the north-west. BR3

TREECREEPER   *Dringwr Bach   Certhia familiaris*
Common and widespread resident in deciduous and mixed woodland. BR3

GOLDEN ORIOLE   *Euryn   Oriolus oriolus*
A scarce passage migrant, mainly occurring in spring.

RED-BACKED SHRIKE *Cigydd Cefngoch* *Lanius collurio*
Rare passage migrant, formerly bred in many parts of Wales before its continuous decline in post-war years.

GREAT GREY SHRIKE *Cigydd Mawr* *L. excubitor*
Rare winter visitor, usually to commons with bracken and scattered trees, or wooded marshland.

JAY *sgrech y Coed* *Garrulus glandarius*
Common resident in conifer and deciduous woods and other places where there are plenty of trees. BR2

MAGPIE *Pioden* *Pica pica*
Abundant species on farms and woodland and anywhere trees are to be found. BR1

CHOUGH *Bran goesgoch* *Pyrrhocorax pyrrhocorax*
Scarce but increasing resident on western coastal cliffs from Holy Island to Gower. Also breeds on mountains in Snowdonia, but apparently no longer in the hills of Mid Wales. BR5

JACKDAW *Jac-y-Do* *Corvus monedula*
Abundant wherever there are sites for nesting in cliffs, quarries, trees, and both derelict and occupied buildings. BR1

ROOK *Ydfran* *C. frugilegus*
Very common but local in some districts, nesting in colonies in sheltered woods and copses. BR1

CARRION CROW *Bran Dyddyn* *C. corone*
Abundant in fields, farms, and wherever there are trees. BR1

HOODED CROW *Bran Lwyd* *C. carone cornix*
Occurs sparingly as a visitor to western Wales, probably from its breeding territories in Ireland or the Isle of Man.

RAVEN *Cigfran* *C. corax*
Quite common in suitable habitats on mountains, moors, sea cliffs, conifer forest, and wooded farmland. Large roosts involving scores or even hundreds of birds may be seen in winter. BR3

STARLING *Drudwen* *Sturnus vulgaris*
Generally a very common nesting species in most urban and rural habitats but can be surprisingly sparse in some country districts. Numerous in flocks in winter. BR2

HOUSE SPARROW *Aderyn y To* *Passer domesticus*
Abundant wherever there is human habitation and buildings for nesting (sometimes nests in trees). BR1

Reed Bunting

TREE SPARROW   *Golfan y Mynydd   Passder montanus*
Local and declining bird, typically of mixed farmland. Commonest in Brecon, Radnor, Carmarthen and Gwent. BR5

CHAFFINCH   *Ji-Binc   Fringilla coelebs*
Probably the most abundant species in Wales, found wherever there are bushes, hedgerows and trees. BR1

BRAMBLING   *Pinc y Mynydd   F. montifringilla*
Rather scarce and local winter visitor. Most often in flocks sometimes mixed with Chaffinches, among beeches, larches and in fields.

GREENFINCH   *Llinos Werdd   Carduelis chloris*
Abundant finch in sheltered shrubby haunts around farms, villages, woods and gardens. Sparser in more open country. BR1

GOLDFINCH   *Nico   C. carduelis*
A very common finch on farmland with rough uncultivated patches, wasteland and uncut hedgerows. Also a familiar bird in gardens. BR2

SISKIN   *Pila Gwyrdd   C. spinus*
Not uncommon resident in upland conifer woods. Has increased steadily since it colonised North Wales in the 1940s and 1950s and now breeds as far south as Glamorgan. BR3. Also winter visitor when it is frequently seen at bird tables. BR3

LINNET   *Llinos   Acanthis. cannabina*
Common resident on gorse clad commons and cliff slopes, rough, weedy fields
and waste ground. Seems commoner near the coast. May be declining. BR2

TWITE   *Llinos y Mynydd   A. flavirostris*
A rare breeding bird to heather moors in North Wales. Occurs elsewhere in
flocks as a scarce winter visitor. BR7

REDPOLL   *Llinos Bengoch   A. flammea*
Resident and winter visitor, breeding in upland plantations and heaths and
boggy patches with birch, alder and willow. May have decreased lately. BR4

CROSSBILL   *Gylfin Groes   Loxia curvirostra*
Breeds spasmodically in upland plantations scattered throughout Wales. Also
winter visitor. BR5

BULLFINCH   *Coch y Berllan   Pyrrhula pyrrhula*
Fairly common resident. Breeds more thinly Caernarfon, Meirionnydd and
Ceredigion. As with most finches, numbers are augmented by winter visitors.
BR3

HAWFINCH   *Gylfinbraff   Coccothraustes coccothraustes*
A very scarce and elusive species, nesting chiefly in Gwent with probably a
few pairs in Radnor and Montgomery. BR7

LAPLAND BUNTING   *Bras y Gogledd   Calcarius lapponicus*
Rare passage migrant or winter visitor, invariably to coastal districts, chiefly
headlands and islands.

SNOW BUNTING   *Bras yr Eira   Plectrophenax nivalis*
A scarce migrant and winter visitor to coasts and uplands, mainly in North Wales.

YELLOWHAMMER   *Melyn yr Eithin   Emberiza citrinella*
Common but declining resident on open farmland with tangled hedgerows,
gorse and bracken slopes, dry heaths and commons. BR3

REED BUNTING   *Bras y Cyrs   E. schoeniclus*
Breeds on marshes, reedy bogs, and margins of lakes, but also occurs in wet
fields with patches of reed. Local in distribution and has declined lately. BR3

CORN BUNTING   *Bras yr Yd   Milaria calandra*
Formerly more widespread, this species now clings to its status as a Welsh
breeding bird on the strength of a few pairs in the Shotton area of Flint. BR7

The following is an additional list of some of the most frequently recorded vagrants and rare visitors to Wales.

Cory's Shearwater
Great Shearwater
Surf Scoter
Night Heron
Wilson's Phalarope
Pectoral Sandpiper
Terek Sandpiper
Buff-breasted Sandpiper
Baird's Sandpiper
Sabine's Gull

Ross's Gull
Gull-billed Tern
Foster's Tern
Short-toed Lark
Richard's Pipit
Tawny Pipit
Bluethroat
Aquatic Warbler
Icterine Warbler
Melodious Warbler

Subalpine Warbler
Barred Warbler
Pallas's Warbler
Yellow-browed Warbler
Woodchat Shrike
Red-breasted Flycatcher
Rose-coloured Starling
Serin
Scarlet Rosefinch
Ortolan Bunting

Honey Buzzard

USEFUL INFORMATION
Addresses

The Welsh Ornithological Society, the Wildlife Trusts, the RSPB and The Welsh Kite Trust all welcome new members, so please write to them for information. Through the British Trust for Ornithology you can participate by contributing data for research while the county bird recorders would like to receive your annual records of birds seen within their areas which will help to build a fuller picture of the status of the various species in Wales. Locally, membership of one of the bird clubs or societies offers the opportunity to share your hobby with like-minded people.

Welsh Ornithological Society
Hon. Secretary	Paul Kenyon, 196 Chester Road, Hartford, Northwich, Cheshire CW8 1LG
Membership Secretary	Dr. D. K. Thomas, Laburnum Cottage, 12 Manselfield Road, Murton, Swansea SA3 3AR

County Bird Recorders
ANGLESEY	S. Cully, Mill House, Penmynydd Road, Menai Bridge, Anglesey LL59 5RT
BRECON	M. F. Peers, Cyffylog, 2 Aberyscir Road, Cradoc, Brecon, Powys LD3 9PB
CAERNARFON	J. Barnes, Fach Goch, Waunfawr, Caernarfon, Gwynedd LL55 4YS
CARMARTHEN	R. O. Hunt, 9 Waun Road, Llanelli, Carmarthenshire SA15 3RS
CEREDIGION	H. Roderick, 32 Prospect Street, Aberystwyth, Ceredigion SY23 1JJ
DENBIGH	N. Hallas, 63, Park Avenue, Wrexham LL12 7AW
FLINT	N. Hallas, 63, Park Avenue, Wrexham LL12 7AW
GLAMORGAN—Two recorders:	
(a) Gower-including Neath, Port Talbot and Swansea.	
	R. H. Taylor, 285 Llangwfelach Road, Brynhyfryd, Swansea SA5 9LB
(b) East Glamorgan	
	S. J. Moon, 36 Rest Bay Close, Porthcawl CF36 3UN
GWENT	C. Jones, 22 Walnut Drive, Caerleon, Newport NP6 1SP
MEIRIONNYDD	D. Smith, 3 Smithfield Lane, Dolgellau, Gwynedd LL40 IBU
MONTGOMERY	B. Holt, Scops Cottage, Pentre Beirdd, Welshpool, Powys SY21 9DL
PEMBROKE	J. W. Donovan, The Burren, Dingle Lane, Crundale, Haverfordwest, Pembs. SA61 1SQ.
	G. H. Rees, 22 Priory Avenue, Haverfordwest, Pembs. SA61 1SQ
RADNOR	P. P. Jennings, Penbont House, Elan Valley, Rhayader, Powys LD6 5HS

Wildlife Trusts
Brecknock Wildlife Trust	Lion House, Bethel Square, Brecon, Powys LD3 7AY
The Wildlife Trust West Wales	7 Market Street, Haverfordwest, Pembrokeshire SA61 1NF
Glamorgan Wildlife Trust	Nature Centre, Fountain Road, Tondu, Bridgend CF31 0EH
Gwent Wildlife Trust	16 White Swan Court, Church Street, Monmouth NP25 3NY
Montgomery Wildlife Trust	Collot House, 20 Severn Street, Welshpool, Powys SY21 7AD
North Wales Wildlife Trust	376 High Street, Bangor, Gwynedd LL57 1YE
Radnorshire Wildlife Trust	Warwick House, High Street, Llandrindod Wells, Powys LD1 6AG

RSPB (Royal Society for the protection of Birds)
Headquarters in Wales	Sutherland House, Castle Bridge, Cowbridge Road East, Cardiff CF1 9AB
North Wales office	Maesyffynnon, Penrhosgarnedd, Bangor LL57 2DW
Headquarters UK	The Lodge, Sandy, Bedfordshire SG19 2DL

The Welsh Kite Trust   The Stable Cottage, Doldowlod, Llandrindod Wells, Powys LD1 6HG

The Wildfowl and Wetlands Trust   Penclacwydd, Llwynhendy Llanelli SA14 9SH

British Trust for Ornithology   The Nunnery, Nunnery Place, Thetford, Norfolk IP24 2PU

Countryside Council for Wales (head office-there are about 15 CCW district offices all told).
Plas Penrhos, Ffordd Penrhos, Bangor, Gwynedd, LL57 2LQ

National Parks
Brecon Beacons National Park   7 Glamorgan Street, Brecon, Powys LD3 7DP
Pembrokeshire Coast National Park   Wynch Lane, Haverford West, Pembs. SA61 1PY
Snowdonia National Park   Penrhyndeudraeth, Gwynedd LL48 6LS

National Trust (head office)   Sgwar y Drindod, Llandudno, Gwynedd LL30 2DE

Welsh Water   Plas y Ffynnon, Cambrian Way, Brecon, Powys LD3 7HP

Forestry Enterprise (Wales)   Victoria Terrace, Aberystwyth, Ceredigion SY23 2DQ

Forest districts
Canolbarth Forest District   Bwlch Nant-yr-Arian, Ponterwyd, Aberystwyth SY23 3AD
Dolgellau Forest District   Government Buildings, Arran Road, Dolgellau, Gwynedd LL40
1LW
Coed-y-cymoedd Forest District   Resolven, Neath SA11 4DR
Llanrwst Forest District   Gwydyr Uchaf, Llanrwst, Gwynedd LL26 0PN
Llanymddyfri Forest District   Llanfair Road, Llandovery, Carmarthenshire SA20 0AL

BIRD CLUBS AND SOCIETIES
No addresses are given since the contact point is often the private address of the secretary of the club at any given time. In the case of the birdwatchers who live in West Wales or Breconshire the Wildlife Trusts are the home bases of the bird clubs or societies. Enquiries can be made locally at libraries, Citizens' Advice Bureaux or Tourist Board offices which may be able to help. The bird clubs and societies however, produce annual or periodic reports on the birds in their areas obtainable from your local recorder (address given above) or other officers.

North Wales   **South Wales**
Cambrian Ornithological Society   Cardiff Naturalists' Society
Clwyd Ornithological Society   Glamorgan Bird Club
Deeside Naturalist Society   Gower Ornithological Society
Wildlife Trust West Wales

Mid Wales
Brecknock Wildlife Trust
Montgomery Field Society
Wildlife Trust West Wales

N.B. The Wildlife Trust West Wales covers the counties of Pembrokeshire, Carmarthen and Ceredigion, and therefore appears in the lists for both Mid and South Wales.

Welsh Bird literature

County avifauna

Where to Watch Birds on Anglesey — (updated guide of R. Hutson 1983), NWWT 1998

Birds of Breconshire — Martin Peers and Michael Shrubb, Brecknock Wildlife Trust 1990

Birds of Caernarfonshire — P. Hope Jones and P. J. Dare, Cambrian Ornithological Society 1976

Birds of Caernarfonshire — John Barnes, Cambrian Ornithological Society 1998

The Birds of Cardiganshire — Ingram G.C.S, H.M..Salmon and W.M. Condry, West Wales Naturalist Trust, 1966

Birds of Denbighshire — P. Hope Jones and J. L. Roberts, Nature in Wales, 1983

Gwent Atlas of Breeding Birds — S. Tyler, Lewis,Venables and Walton, Gwent Ornithological Society, 1987

An Atlas of Breeding Birds in West Glamorgan — Thomas, Derek K. Gower Ornithological Society 1992

Birds of Glamorgan — Clive Hurford and P. Lansdown, Cardiff Nat. Soc. 1995

A Guide to Gower Birds — Grenfell, Harold and Thomas, Gower Ornithological Society 1982

Birds of Gwent — Ed. P. N. Ferns et al, Gwent Ornithological Society, 1977

Birds of Merioneth — P. Hope Jones, Cambrian Ornithological Society, Colwyn Bay, 1974

Montgomeryshire Bird Report 1993 & 1994 — Holt (ed).

Birds in Pembrokeshire — Huw Morgan, Rosedale Publications, 1996

Birds of Pembrokeshire — Jack Donovan and G.H. Rees, Dyfed Wildlife Trust, 1994

Birds of Radnorshire and mid Powys — M. Peers 1985

The Birds of Bardsey — P. Roberts, Bardsey Bird and Field Observatory, 1985

Birds of Skokholm — Michael Betts, Dyfed Wildlife Trust, 1991

N.B.   There is no Carmarthenshire avifauna since Ingram and Salmon's *Handlist of the Birds of Carmarthenshire 1954*, but there is a good annual report, *Carmarthen Birds*.

General

The Red Kites of Wales — A. V. Cross & P. E. Davis, The Welsh Kite Trust 1998

In Search of Birds in Mid Wales — Brian O'Shea and John Green, 2nd ed. 1991

Silent Fields — Roger Lovegrove et al, RSPB Wales, 1995

The Natural History of Wales — William M. Condry, Collins, 2nd ed. 1990

Birds in Wales — Roger Lovegrove, Poyser, 1994

Where to Watch Birds in Wales — David Saunders, 2nd ed. 1992

N. B. This is not a complete list.

Care in the Countryside

Remember the Country Code!

Enjoy the countryside and respect its life and work	*Leave livestock, crops and machinery alone*
Guard against all risk of fire	*Take your litter home*
Fasten all gates	*Help to keep all water clean*
Keep your dog under close control	*Protect wildlife, plants and trees*
Keep to public paths across farmland	*Take special care on country roads*
Use gates and stiles to cross fences, hedges and walls	*Make no unnecessary noise*

~~~~~~~~~~

Wherever you are, it is important to avoid disturbing birds, particularly when they are nesting. Even though your intention is only to observe and enjoy them, birds will regard you as a predator and keep their distance. A bird disturbed from its nest may be reluctant to return and its eggs may go cold and become infertile. Unfed nestlings will squawk or the commotion created by the adults will soon attract the sharp eyes and keen instincts of real predators. Disturbed vegetation close to occupied nests will also be quickly noticed by carnivores with highly developed senses. A startled bird put off the nest will often desert its eggs or even its young.

If you are vigilant and aware, you can enjoy the birds without putting them at risk. If a bird seems agitated, move quietly and quickly from the immediate area. The same applies if you are sitting for a while in the same place and a single bird or a pair remain close or keep returning to your spot. It is unlikely that they are just being sociable, and unless they are coming for food at an established picnic site, they are most likely to have a nest in the vicinity! Large birds of prey like the Kite will often circle overhead if you are too close to their nests. The months of April, May and June, and to a lesser extent March and July, are the critical months for nesting birds. The 1981 Wildlife and Countryside Act makes it an offence to disturb rare birds at the nest (without a licence usually given for research purposes). It is also illegal to take the eggs or young of ANY species except in very special authorised circumstances (related to conservation or farming activities).

Outside the breeding season the risks to birds arising from disturbance are generally fewer but do be careful to avoid putting roosting birds to flight, or those tired migrants resting on the beaches, estuaries and headlands.

INDEX OF BIRDS

(Plover)
Golden Plover   58,59,97,98
Grey Plover   49,97,98
Little Ringed Plover   77,78,92,103
Ringed Plover   43,92,97
(Pochard)
Pochard   82,83,96
Red-crested Pochard   121
Puffin   46
Quail   36
(Rail)
Water Rail   81,83
Raven   51,55,99
Razorbill   46
Redpoll   66,67,99
(Redshank)
Redshank   49,81,89,90,92
Spotted Redshank   49,92
(Redstart)
Black Redstart   100
Redstart   29,62,
Redwing   31,99
Robin   29,72
Rook   30,99
(Rosefinch)
Scarlet Rosefinch   52
Ruff   83,90
Sanderling   97,103
(Sandpiper)
Baird's Sandpiper   138
Buff-breasted Sandpiper 138
Common Sandpiper   74,92
Curlew Sandpiper   50,86
Green Sandpiper   49,92
Marsh Sandpiper   86
Pectoral Sandpiper   138
Purple Sandpiper   98
Terek Sandpiper   86,
White-rumped Sandpiper 86
Wood Sandpiper   49,
Scaup   82,95
(Scoter)
Common Scoter   48,49,95
Surf Scoter   138,
Velvet Scoter   95
Serin   138
Shag   48

(Shearwater)
Cory's Shearwater   47
Great Shearwater   47
Manx Shearwater   46,47
Mediterranean Shearwater   47
Sooty Shearwater   47
Shelduck   49
Shoveler   81,83
(Shrike)
Great Grey Shrike   100
Red-backed Shrike   37,38
Woodchat Shrike   52
Siskin   66,67,99
(Skua)
Arctic Skua   48,93
Great Skua   48,93
Long-tailed Skua   48,93
Pomarine Skua   48,93
Skylark   37,58
Smew   95
(Snipe)
Jack Snipe   98
Snipe   58,59,89,98
(Sparrow)
House Sparrow   29,79
Tree Sparrow   29,37,79
Sparrowhawk   28,31,68,108,109,110
Spoonbill   87
(Starling)
Rose-coloured Starling   52
Starling   30,99
(Stint)
Little Stint   49,86
Stonechat   50,53
(Stork)
Black Stork   87
White Stork   87
Swallow   31
(Swan)
Bewick's Swan   94
Mute Swan   90,94
Whooper Swan   94
Swift   113,130
Teal   59,81,96
(Tern)
Arctic Tern   42,93
Black Tern   42,80,81
Common Tern   42,93

Forster's Tern	138	Icterine Warbler	52
Gull-billed Tern	138	Marsh Warbler	85
Little Tern	42,43	Melodious Warbler	52
Roseate Tern	42	Pallas's Warbler	52
Sandwich Tern	42,93	Reed Warbler	83,84,
(Thrush)		Savi's Warbler	84
Song Thrush	29,68,99	Sedge Warbler	84
Mistle Thrush	31,99	Subalpine Warbler	52
(Tit)		Willow Warbler	31,32,62,68
Bearded Tit	84,87,98,99	Wood Warbler	29,65
Blue Tit	29,68	Yellow-browed Warbler	52
Coal Tit	29,32,67,68,99	Waxwing	101
Great Tit	29,68	Wheatear	51,59,102
Long-tailed Tit	33,99	Whimbrel	52,101,102
Marsh Tit	32,33,79	Whinchat	61,83
Penduline Tit	98,99,	**(Whitethroat)**	
Willow Tit	32,33,79	Lesser Whitethroat	36
Tree Creeper	33	Whitethroat	36,50,51
Turnstone	98	**(Wigeon)**	
Twite	61,100	American Wigeon	106
(Wagtail)		Wigeon	96
Grey Wagtail	73,74	Woodcock	68,98
Pied Wagtail	30,74	Woodlark	38,65
Yellow Wagtail	78,79	**(Woodpecker)**	
(Warbler)		Great Spotted Woodpecker	33,34
Aquatic Warbler	85	Green Woodpecker	33,34
Barred Warbler	52,	Lesser Spotted Woodpecker	34,79
Cetti's Warbler	85	Wood Pigeon	31,69
Dartford Warbler	85	Wren	30
Garden Warbler	32,68	Wryneck	130
Grasshopper Warbler	68,83,85	Yellowhammer	35,37,62,68

Note

(1) The generic names are in **bold** print simply to make it easier to use the index which applies mostly to the descriptive chapters one to eight. For further reference to species, see birds listed under each location of the three regions and also the systematic check list beginning on page 118.

Sparrowhawk